AF707409

sipri

Stockholm International Peace Research Institute

SIPRI is an independent institute for research into problems of peace and conflict, especially those of disarmament and arms regulation. It was established in 1966 to commemorate Sweden's 150 years of unbroken peace.

The Institute is financed by the Swedish Parliament. The staff, the Governing Board and the Scientific Council are international.

The Board and Scientific Council are not responsible for the views expressed in the publications of the Institute.

Stockholm International Peace Research Institute
Pipers väg 28, S-171 73 Solna, Sweden
Cable: Peaceresearch, Stockholm
Telephone: 08-55 97 00

Cover design: *The scorpion is an arachnid with a long, slender, jointed tail ending in a curved, poisonous sting. As an animal that is said to even sting and poison itself, it is analogous to mankind's use of chemical and biological weapons, which easily contaminate the users themselves.*

The Chemical Industry and the Projected Chemical Weapons Convention

The Chemical Industry and the Projected Chemical Weapons Convention

Volume I

Proceedings of a SIPRI/Pugwash Conference

sipri

Published on behalf of
Stockholm International Peace Research Institute
by
Oxford University Press

1986

This book has been printed digitally and produced in a standard specification in order to ensure its continuing availability

OXFORD
UNIVERSITY PRESS

Great Clarendon Street, Oxford OX2 6DP

Oxford University Press is a department of the University of Oxford.
It furthers the University's objective of excellence in research, scholarship, and education by publishing worldwide in

Oxford New York

Auckland Cape Town Dar es Salaam Hong Kong Karachi
Kuala Lumpur Madrid Melbourne Mexico City Nairobi
New Delhi Shanghai Taipei Toronto

With offices in

Argentina Austria Brazil Chile Czech Republic France Greece
Guatemala Hungary Italy Japan South Korea Poland Portugal
Singapore Switzerland Thailand Turkey Ukraine Vietnam

Oxford is a registered trade mark of Oxford University Press in the UK and in certain other countries

Published in the United States by Oxford University Press Inc., New York

Reprinted 2008

ISBN 978-0-19-829107-7

The Chemical Industry and the Projected Chemical Weapons Convention

Volume I

Proceedings of a SIPRI/Pugwash Conference

sipri

Published on behalf of
Stockholm International Peace Research Institute
by
Oxford University Press
1986

This book has been printed digitally and produced in a standard specification in order to ensure its continuing availability

OXFORD
UNIVERSITY PRESS

Great Clarendon Street, Oxford OX2 6DP
Oxford University Press is a department of the University of Oxford.
It furthers the University's objective of excellence in research, scholarship,
and education by publishing worldwide in
Oxford New York
Auckland Cape Town Dar es Salaam Hong Kong Karachi
Kuala Lumpur Madrid Melbourne Mexico City Nairobi
New Delhi Shanghai Taipei Toronto
With offices in
Argentina Austria Brazil Chile Czech Republic France Greece
Guatemala Hungary Italy Japan South Korea Poland Portugal
Singapore Switzerland Thailand Turkey Ukraine Vietnam

Oxford is a registered trade mark of Oxford University Press
in the UK and in certain other countries

Published in the United States
by Oxford University Press Inc., New York

Reprinted 2008

ISBN 978-0-19-829107-7

PREFACE

During 24-26 October 1985, 30 specialists from 18 countries met privately at the Royal Academy of Sciences in Stockholm to confer on problems which the chemical industry and the projected Chemical Weapons Convention may present for one another. The conference was unpublicized, and is reported here, with the consent of the participants, for the first time. It was convened jointly by the Stockholm International Peace Research Institute (SIPRI) and the Pugwash Conferences on Science and World Affairs. All those involved are identified in the Participation List, below.

The form of the present publication is set by our wish to disseminate the knowledge gained during the conference as rapidly as possible--above all, to make it available to other bodies now embarking upon work within the same general area. Readers will find two main things: a selection of the papers presented at the conference, many of them subsequently revised, and (in Volume 1) a summary report of the discussions that took place. The latter follows the normal practice of Pugwash specialist meetings. It does not identify who expressed what view, nor is it a consensus report in the sense that all participants have read and approved it. Rather, it sets out personal impressions of the proceedings that were formed by two of the participants, one from the German Democratic Republic, and the other from the United Kingdom. These rapporteurs were assisted in their joint task by a member of the SIPRI research staff serving with the secretariat. A tape-recording had been made of the entire proceedings.

Purpose of the conference

The stimulus to convene the conference had come from developments at the CD--the 40-nation Conference on Disarmament in Geneva. There, negotiations for a comprehensive chemical-warfare (CW) disarmament treaty--the projected Chemical Weapons Convention--had commenced in 1984 after some 15 years of preliminary talks and elaboration. The early exploratory work within the CD and its antecedent bodies had made it clear that the particular issue of verifying nonproduction of CW weapons would be among the more difficult ones for the negotiators to resolve; and, at this writing, they have still not addressed it in a concerted fashion. Both SIPRI and Pugwash had done much work on the issue during 1968-79, so there was point in them returning to it now that actual negotiation seemed imminent.

The CD is trying to negotiate a treaty that will establish an international regime of CW disarmament. There will be no agreement unless the terms of the treaty can give prospective states parties confidence that their security is likely to be served better by their joining the regime than by their remaining outside it. For that assurance, the regime must incorporate national and international measures of verification that can function, not only as a monitor of compliance, but also--at least as important--as a means for dispelling uncertainties about the real nature of activities that might otherwise be mistaken for prohibited ones.

There are certain areas of normal and unexceptionable peacetime activity which could display such ambiguity. Chemical industry is one. It is a feature of CW armament that much of the production capacity, the skills and the supplies of toxicants, or of intermediates for toxicants, which it needs may exist even in an industry that is serving civilian needs only. This circumstance has in the past prompted erroneous suspicions that particular chemical firms have been engaged in CW armament, either for their own or for other governments; and it will no doubt continue to do so. The projected Chemical Weapons Convention could protect the chemical industry against harmful but unfounded accusations of this type. But that it could do only if the regime it created laid down procedures which allowed the industry to demonstrate that its resources were not being diverted or exploited for purposes outlawed by the treaty. Such arrangements would in turn protect states from unjustified charges of noncompliance with the treaty. Conversely, if the industry found itself unable to accept the controls (for that is what the requisite procedures would be), there could be no treaty and therefore no protection against scurrilous or malicious aspersion. Access to the wider benefits that the treaty could bring to the national security would also be blocked.

So, for governments to be able to negotiate an international treaty which provided adequate assurance, they would need to be aware of any difficulties which a proposed regime might present to the chemical industry, their own and that in other countries. And, for their part, national chemical industries would need to be sufficiently informed, both about the range of available options for the regime, and about those of their activities which might be considered dangerously ambiguous from a CW standpoint.

To Pugwash and SIPRI, these appeared to be matters on which the existing channels of communication linking the negotiators into the relevant areas of expertise and interest might well be inadequate, especially the transnational channels of communication. Any inadequacy would be bound to prolong the time taken to resolve the issue of non-production verification, once it had finally been taken up by the negotiators; and this could make the forward momentum of the negotiations unnecessarily difficult to sustain. They felt, accordingly, that their respective experiences, not least in transnational communication, could be put to useful purpose in augmenting the CD's own channels.

It was this, then, that was the chief purpose of the conference. The means chosen was to provide occasion for extended exchanges of view between people from different chemical industries, from CW laboratories and from the Geneva negotiating effort within an environment that was both private and noncommittal. The objective was not to seek consensus among these people on any particular position or recommendation. It was simply to be an exercise in mutual understanding: a preliminary exploration of problems for which solutions may need to be found.

Scope of the conference agenda

It is the nature of the relationship between government and industry in a particular country that will mould the form of those assurance procedures which the domestic chemical industry may regard as an acceptable degree of control. And how that relationship is perceived in other countries will influence the confidence existing abroad in the adequacy of the controls. Differences in the relationship from one country to another fall into three broad categories, each one presenting a different facet of the overall compliance-assurance requirements. As outlined below, they indicate the spread of the agenda put before the conference.

Market economies. In economies of the type that characterise the OECD countries, the degree to which the chemical industry is independent of government may mean that the latter could face substantial difficulty in achieving domestic consensus on controls that are both tolerable by the industry and adequate for the purposes of the projected treaty regime. Moreover, within this category, the degree of independence of the chemical industry varies appreciably from country to country. This may further impede consensus by increasing sensitivity to the possibility that a particular set of controls that might form part of the regime could impact asymmetrically on the international competitiveness of each national industry. The controls might involve international on-site inspections of civil-industrial facilities, for example, or obligations to report production and consumption data for some particular chemicals. If, therefore, a satisfactory treaty regime is to be developed, the negotiators need to understand why the chemical industry - their own and that of other countries - might prefer to preserve the privacy of particular activities which conceivable controls might invade. And the industry needs to understand why the continued privacy of such activities may be incompatible with an adequate treaty regime.

Centrally planned economies. In economies of this type, such as are found in the Socialist countries, decisions on the feasibility and acceptability of controls will rest with the central authorities, which is to say with government. The problems then faced by government in achieving domestic consensus on such decisions are therefore likely to be of a different and perhaps easier kind than those facing government in the market-economy countries. But there is an additional difficulty which this group of states must somehow overcome: that of projecting assurance internationally that their chemical industries are indeed observing the procedures. A prerequisite for this is a wider international understanding of the mechanics of the relationship between government and chemical industry in the centrally planned economies: the manner in which the latter is in practice regulated and controlled by the former, and the character assumed by proprietary information. For example, are there conditions found only within centrally planned economies that might militate against controls which the chemical industries of market-economy countries might, in contrast, find acceptable?

Developing countries. As compared with the fully industrialised countries, chemical industry in the developing countries tends to command a favoured position by virtue of the recognition accorded at the political level to its central role in furthering national development goals. Any controls required under the projected Chemical Weapons Convention are therefore liable to be perceived as contrary to the national interest because of the constraints they might be thought to place on the overall development process. Whether the benefit to the national security which an effective Chemical Weapons Convention could bring to such countries would in fact, because of its associated control regime, constitute a cost to development requires more study than it yet appears to have received. A prerequisite is a clearer and wider understanding of the mechanisms whereby possible regimes could conceivably inhibit industrial development. Is the dominant consideration an international one--that of, for example, the competitiveness of industrial products on the world market? Or does it lie, rather, in domestic industrial infrastructure? And would the sorts of controls involved really be any more onerous in their impact on development than those necessitated by the requirements of industrial hygiene, environmental protection, even public safety?

Organisation of the conference

Specialist participation in the conference was secured primarily through the good offices of the Pugwash Chemical Warfare Study Group (PCWSG). This body dates back to 1973 when it was established in order to provide opportunity for expert but unofficial East-West contacts on CW arms control that would complement the formal intergovernmental channels. Membership of the group is not limited to natural scientists: it includes military people and governmental officials too, all of them having an expertise in one or another aspect of CW preparedness. The group has met on average rather more than once a year, typically with a 2 - 4 day prepared agenda structured according to matters of current topicality. Experts from 27 countries have now participated in these meetings: Australia, Belgium, Brazil, Canada, China, Czechoslovakia, Denmark, Egypt, Finland, France, the two German states, Hungary, India, Israel, Italy, Japan, the Netherlands, Norway, Poland, Romania, Sweden, Switzerland, the United Kingdom, the USA, the USSR and Yugoslavia.

The PCWSG has long enjoyed a close and mutually supportive relationship with SIPRI, as did its antecedent Biological Warfare Study Group. Its first and seventh workshops, in 1974 and 1979 respectively, were organised as joint SIPRI/Pugwash meetings, SIPRI publishing the proceedings.[1,2] It was on this model that the conference, in effect the eleventh PCWSG workshop, was organised.

The conference was not the first occasion on which either SIPRI or the PCWSG has studied chemical-industry/chemical-disarmament relationships. Much detailed work was done by SIPRI during 1970-72, chiefly by convening an East-West expert working party on the idea of phosphorus-accountancy as a means of verifying nonproduction of nerve gas, and by organising symposia on this and related techniques. That work is published.[3,4] In addition, the subject of nonproduction verification has frequently been on the agenda of PCWSG meetings, most notably the 1977, 1978 and 1979 workshops during which the group was invited to visit industrial establishments in, respectively, the FRG,[5] the USA[6] and Sweden,[2,7] the management of the plants visited participating in the workshops. This model has since been copied at the governmental level by several CD delegations.

It was SIPRI, however, which, with assistance from the Swedish Pugwash Group, the Pugwash Executive Office in Geneva, and the Science Policy Research Unit of the University of Sussex, England, bore the main financial and administrative burdens of the present conference. A debt of gratitude is owed, too, to the Royal Swedish Academy of Sciences for making its fine premises and facilities available for the conference. The welcoming address by its foreign secretary, Olof G. Tandberg, set a tone for the proceedings which the spirit of dispassionate inquiry that seemed to emanate from the portraits of Swedish men of science covering the walls of the conference chamber further enhanced. It seemed especially fitting that we should be working under the gaze of Carl von Linné and Jöns Jacob Berzelius.

Literature Cited

1. SIPRI, Chemical Disarmament: New Weapons for Old (Stockholm: Almqvist & Wicksell, 1975)

2. SIPRI, Chemical Weapons: Destruction and Conversion (London: Taylor & Francis, 1980)

3. SIPRI, The Problem of Chemical and Biological Warfare, Volume 6: Technical Aspects of Early Warning and Verification (Stockholm: Almqvist & Wicksell, 1975), pp 81-88, 164-294.

4. SIPRI, Chemical Disarmament: Some Problems of Verification (Stockholm: Almqvist & Wicksell, 1973)

5. Pugwash Newsletter Vol 15 No 3 (January 1978), pp 84-95

6. Pugwash Newsletter Vol 16 No 1 (July 1978), pp 4-12

7. Pugwash Newsletter Vol 17 Nos 1-2 (July-October 1979), pp 40-48.

Julian Perry Robinson
Convenor and Series Editor

Abstract

The Chemical Industry and the Projected Chemical Weapons Convention: Proceedings of a SIPRI/Pugwash Conference.
Oxford University Press, Oxford, 1986, 147 pp. (Stockholm International Peace Research Institute). ISBN 0 19 829107 8.

This publication, which is printed in 2 volumes, presents a selection of papers from and a report on a conference on the topic 'The Chemical Industry and the Proposed Chemical Weapons Convention' which was held in Stockholm, Sweden during 24–26 October 1985. It was convened jointly by the Stockholm International Peace Research Institute (SIPRI) and the Pugwash Conferences on Science and World Affairs. Thirty specialists from 18 countries met privately to confer on the problems associated with a comprehensive chemical-warfare disarmament treaty. Volume One, CBW Studies No. 4, contains a summary report of the discussions written by two of the participants, one from the GDR and the other from the UK, as well as 3 lengthy background papers related to the topic of the conference. Volume Two, CBW Studies No. 5, contains 16 papers selected from among those presented at the conference. Both volumes are illustrated with numerous tables, figures and other relevant documents.

CONTENTS--VOLUME ONE

CONTENTS--VOLUME TWO

PARTICIPATION LIST

Institutional affiliations are given here for information only and do not imply participation by the people concerned in anything other than their private capacities.

Akiyama, Dr I	Mitsubishi Research Institute, Inc, Tokyo, Japan
Babievsky, Prof K	Institute of Organo-Element Compounds, USSR Academy of Sciences
Blackaby, Mr F	Director, SIPRI
Carpenter, Dr W D	Monsanto Company, USA
Cooper, Dr G H	Chemical Defence Establishment, Porton Down, UK
Ezz, Maj-Gen Dr E	Formerly Head of Scientific Research Branch, Egyptian Armed Forces
Fokin, Acad A V	Praesidium, USSR Academy of Sciences
Goeres, Prof Dr E	Director, Institute of Pharmacological Research, Berlin, GDR
Hamblin, Dr M C	Director, Defence Sciences Division, Defence Research Establishment Suffield, Canada
Hoffmann, Prof Dr H	Chemical Research Director, Agrochemical Division, Bayer AG, FRG
Jacchia, Prof E	Director-General (hon), European Community; and Director, Centro di Studi Strategici, Rome, Italy
Kaplan, Dr M M	Secretary-General, Pugwash Conferences on Science and World Affairs
Lau, Dr A	Senior Research Officer, National Defence Research Institute, Sweden
Lazary-Teixeira, Dr C A	Adjunto Secretaria-General, Conselho de Seguranca Nacional, Presidencia da Republica, Brazil

Lohs, Prof Dr K	Academy of Sciences, GDR
Lundin, Dr S J	Director of Research, National Defence Research Institute, Sweden
Mate, Dr L	Military Medical Research Centre, Budapest, Hungary
Meselson, Prof M S	Department of Biochemistry and Molecular Biology, Harvard University, USA
Miettinen, Prof J K	Director, Finnish Research Project on the Verification of Chemical Disarmament
Mikulak, Dr R	Senior Scientist, US Arms Control & Disarmament Agency
Ooms, Dr A J J	Director, Prins Maurits Laboratory TNO; member of the Interdepartmental Advisory Council on Peace and Security; and member of the Netherlands Delegation to the CD
Ramachandran, Dr P K	Director, Defence R&D Establishment, Gwalior, India
Rautio, Dr M	Coordinator, Finnish Research Project on the Verification of Chemical Disarmament
Robinson, Mr J P P	Senior Fellow, Science Policy Research Unit, University of Sussex, UK
Rump, Prof S	Military Institute of Hygiene & Epidemiology, Warsaw, Poland
Santesson, Dr J	Head of Chemistry Division, NBC Protection Department, National Defence Research Institute (FOA 4), Sweden
Trapp, Dr R	SIPRI; and Forschungsstelle f. Chemische Toxikologie, Leipzig, GDR
Vincken, Mr J	Shell Nederland Chemie BV, Netherlands
Vojvodic, Maj-Gen Prof V	Military Medical Academy, Belgrade, Yugoslavia
Åberg, Prof B	Chairman, Swedish Pugwash Group

Observers

Björnerstedt, Dr R	Chairman, SIPRI Governing Board
Carlsson, Amb B	Ministry for Foreign Affairs, Stockholm, Sweden

Nemec, Ing T	SIPRI
Taylor, Mr J L	Embassy of the United Kingdom, Stockholm
Zhu, Mr W Y	Embassy of the People's Republic of China, Stockholm

Secretariat

de Bellmond, Ms A	SIPRI
Gilligan Borg, Ms J	SIPRI
Helleday, Ms A	SIPRI
Kjellgren, Ms B	SIPRI
Mårtensson, Ms K	SIPRI
Vargas-Fuentes, Dr R	SIPRI
Winqvist, Ms G	SIPRI

The conference Chairman was Dr Johan Lundin. The Head of Secretariat was Ms Gun Winqvist.

SUMMARY REPORT OF THE PROCEEDINGS

J. P. Perry Robinson and Ralf Trapp, with Kristina Mårtensson

The status of our Summary Report is described in the Preface to this volume. We present it in two parts. The first is an overview of the main features of the conference. The second is a more detailed narrative account. We make no attempt to summarize the content of the papers presented. Our purpose is to convey an impression of the main directions taken by the discussions during the two and a half working days of the conference.

I Overview

The premise on which the conference was founded was that normal workings of civilian chemical industry cannot remain entirely unaffected by any chemical-warfare (CW) disarmament convention that governments may agree to join. Yet some delegations at the Conference on Disarmament (the CD) in Geneva have thus far refrained from endorsing that proposition wholeheartedly, at least insofar as it implies that control procedures may need to be applied to the industry. And even for CD delegations which accept the implication, there is the further dilemma posed by the manner in which the control procedures should best be applied, above all the essentially polictical question of the degree to which there should be an international element (as well as the nature of that element) in whatever schemes are to be imposed upon civilian chemical industry. By the end of the conference we were left in no doubt that the projected treaty regime would indeed have to provide for some form of control on civilian chemical industry - control in the senses both of monitoring and of regulation; and we had gained a rather clearer idea than we had had before of the possible shape of those controls. But the question of the international element--the composition, functions and authority of the international control organ that might, for example, be established under the auspices of a consultative committee of status parties--remained unresolved.

Civilian chemical industry could in theory become involved with CW armament in several different ways. For example, its resources could be drawn upon for the requisite research and development (R & D); it could provide skills and manpower needed to design and construct production facilities; and the operation of those facilities, whether they were or were not isolated within some military sector of the economy, might be run by civilian firms. Further, supplies of the requisite raw materials and production intermediates might be acquired by diversion from civilian industry or else by importation from the industries of other countries. The purpose then, of any industry controls established by the CWC would be to provide assurance internationally that such diversions were not taking place. So what the conference was primarily considering was the central aspect of what the CD refers to as 'nonproduction verification' (NPV): measures whereby assurance can be gained that chemicals are not being produced in quantities and for purposes other than those which are to be permitted under the treaty.

The processing of chemicals on any scale within civilian manufacturing industry, and subsequent commerce in them, are regulated by governmental agencies in all industrialised countries. Just how, and for what purposes were the subject of several of the conference papers; and further illustrations of the nature and extent of these national controls were provided to us in other forms, most notably as regards Brazil,[1] the GDR[2] and the USA[3]. An outstanding feature of the conference was the readiness of the industry participants to speak to the questions of whether and how an effective NPV regime might be developed from this body of national regulatory practice, having regard to the impacts such a regime might have on the normal workings of their industry.

The discussion was at its most detailed with regard to the developed countries of East and West, especially the Federal Republic of Germany, the Netherlands and the USA; but it also extended to the situation in the developing countries. It is particularly pleasing to be able to report the discussions which took place regarding NPV and the chemical industry of India.

Some of the ground covered at the conference had of course also been studied in prior work, so to this extent our account of the proceedings adds nothing especially new. But the conference also covered ground which has not, to our knowledge, been studied in any comparable detail before. It would be useful here, in this summary overview, to encapsulate what was said on such topics. But in doing so we would risk implying a consensus of view which, although it might well have existed, it was no purpose of the conference to seek. So we will simply flag here those particular features of our narrative report below which seem to us to add significantly to the existing literature on NPV. There were five main ones:

(a) The attention given to the concept of 'negative proof'. In its application, especially to the process-sampling that might be undertaken by inspectors at industrial facilities, the concept could become an effective device for minimizing jeopardy to intellectual property and commercial confidentiality.

(b) The attention given to the difficult problems presented by the operations of commodity chemical brokers and by other dealers and traders in chemicals.

(c) The discussion of the importance of striking a balance between the scope of the projected control regime and the size of its administrative organs, whether national or international. An over-ambitious control regime might risk bringing into existence a large, and therefore potentially counterproductive, bureaucracy as well as generating further technical problems.

(d) The discussion of the concept of small-scale 'permitted production' facilities. In the course of the Geneva negotiations, this concept has found expression in different forms, all of them, however, aimed at the same objective: that of bringing under the most stringent control all legitimate production of those chemicals whose properties--supertoxicity, for example--might make them exceptionally amenable to chemical-warfare purposes. We were led to conclude from this discussion that it would be very difficult for some parts of the world chemical industry to accept a concept in which production within any one country of, say, a pharmaceutical falling within the control category had to be undertaken in the same facility as production of the small quantities of nerve gas (for example) required for work on antichemical protection.

(e) The attention given to the skills and personal qualities that should be sought in people to be tasked with inspection duties.

Perhaps the most constructive overall conclusion which it is possible for us record here is this. The regulatory practices under which chemical industry operates vary in points of detail and administration from country to country. Yet, considered as a foundation upon which an NPV regime might be rested by the projected Chemical Weapons Convention, it was the similarities of the national regulatory regimes, not their differences, which were the more striking. Of course, this was an impression gained from what we heard about only a limited number of countries. Yet those countries appeared to be broadly representative of the East as well as the West, the South as well as the North. We did not learn exactly how the regulatory regimes could best be extended so as to serve NPV purposes as well; that would have to wait upon those political decisions referred to above about the balance that was to be struck between the national and the international elements of the overall scheme. But what we did learn was that the information--verifiable information--that currently flows from chemical industry to government includes a high proportion of the information that would seem to be needed for an adequate NPV system. Such additional controls as might be required seem unlikely, therefore, to constitute an intolerable burden for the chemical industry. They may need to be no more than a minor expansion of controls already in place. We heard no dissent from the view that such an approach to NPV was a sensible one. On the contrary, the approach was endorsed; and we heard from industry participants much constructive comment on the various pitfalls that it might encounter and how they could be avoided.

II Narrative Report

The aims, organisation and sponsorship of the conference are described in the Preface to the present volume. The Preface also quotes from the basic information paper originally sent to potential participants outlining the sorts of topics on which they were invited to present papers. Eighteen of the participants duly prepared such papers; two others provided background materials. The conference also had before it three background papers, two of them prepared by the authors of the present report, the third commissioned for the conference from the Industrial Chemistry Group at the University of Surrey[4]. The three background papers, somewhat revised, are reproduced later in this volume; other papers presented are to be found in Volume 2.

The Chairman of the conference, Dr. Johan Lundin, had proposed the following agenda:

(1) Regulatory practices in national chemical industries.

(2) Chemicals and chemical production requiring regulation under the projected Chemical Weapons Convention.

(3) Possible implications for the chemical industry.

The intention was that we would begin by hearing from industry participants about the nature of the regulatory practices under which their firms operated in their respective countries. Those regimes might, it was expected, have features which could provide a framework for whatever control regime might be needed for the projected Chemical Weapons Convention (CWC). Under the second item of the agenda, we would then consider the particular chemicals and industrial activities which might need to be brought within that control regime. Finally, under the last item of the agenda, we would consider how such a regime might actually operate, having regard to the various impacts it might have on industrial practices and vice versa. That was indeed how the conference proceeded. The submitted papers grouped themselves rather evenly under the three agenda items. Their authors were asked to present oral summaries which then served as the point of departure for discussions under each head.

These discussions we now summarize. Our narrative generally follows the order of presentation, but in some places we have thought it useful to provide a more analytical account.

II.1 Regulatory practices in national chemical industries

We heard from several participants about the controls and regulations affecting the chemical industry in their countries. The legal and technical structure of these regulatory practices, though it varied considerably from country to country, meant that the chemical sector of industrialized economies was one of the most stringently controlled of all industries. This was so for the market economies and for the centrally planned economies of the Socialist states. There were historical reasons for this situation, and reasons originating in the particularities of chemical industry. The purposes served included: safety in the workplace; public safety; environmental protection, including air- and water-pollution control, and (as regards waste disposal) resource conservation; the monitoring by central government of the overall economic performance of the chemical industry and of related economic parameters; and such limiting of imports or exports as was required to implement prevailing governmental policies.

Industrialized countries

In the USA, chemical corporations are required to report a substantial body of data, both on a regular basis and on an *ad hoc* basis, to several state and federal authorities. A sizeable fraction of employees spend much of their time generating and processing these data: we were told of one chemical corporation where, of the 40,000-50,000 people employed, some 1700 spend their time satisfying the requirements of regulatory agencies. The information that is reported by the corporations to state and federal government is in the form of material statistics (quantitative data on raw materials, processed materials and end products), qualitative information about certain of the properties of the materials involved, including toxicological information, statistics on the labour situation at each processing facility, as well as statistical and analytical data on air and liquid effluents and solid wastes. Parts of this information are regarded as confidential by the corporations, mindful of their competitiveness. Provisions exist within the regulatory legislation for protecting that confidentiality, requiring the governmental agencies to which the information is to be reported to protect it from unauthorized access or publication. The information reported enables the regulatory agencies to form a detailed picture of plant operations. They are empowered to verify it by means that include on-site inspections and the collection of samples (of waste water, for example) for analysis.

Regulatory practices in other developed market economies are, or are now becoming, broadly similar in extent and degree of control to those of the USA, which are themselves, through implementation of the Toxic Substances Control Act, becoming still more stringent. In the Netherlands, for example, where there is tight legislation on environmental protection, work-place safety and waste disposal, a Chemical Substance Act will soon become law (reflecting a common policy adopted by the European Communities) that will require notification of the production or marketing of any new substance in an amount exceeding one tonne per year. As in the USA, the volume of reported data is substantial. Although some of that information is provided by industry on a confidential basis, a lot is in fact accessible to the general public. It was estimated that about 80 per cent of the data needed to satisfy the NPV scheme which Britain had proposed at the CD[5] was in fact already reported to the Netherlands authorities in accordance with existing regulatory practices.

Chemical-industry controls have broadly similar objectives in all the market-economy countries, but there are some significant differences in the manner in which they have been constructed. Thus, in Japan, the control mechanisms depend entirely upon listings of the individual chemicals that are to be subject to monitoring. Generic identification, as by the specification of broad classes of chemical (on, say, toxicological criteria) without the need to identify each and every member of the class, is not used. The purview of each regulatory agency is defined by such lists. This would mean that, if the rules and procedures laid down by the projected Chemical Weapons Convention were couched in terms of groups of chemicals, Japanese authorities would be required to draw up their own lists of individual chemicals before domestic implementation of the convention within the Japanese chemical industry could become possible.

Regarding the Socialist countries, we were told in most detail of regulatory practices in the GDR, where some 340,000 people are employed in the chemical industry. These practices display a stringency comparable to ones in the market economies. The key legislation is the Poisons Act, which extends to the manufacture, trade and use of all chemicals in the GDR. Specially empowered organisations enforce this law, the control functions falling to the Ministry of Health, special police departments and the trades unions in particular. These control organs provide surveillance both of the regulated materials and of the personnel entitled to handle them. There are comparable regulatory mechanisms in the USSR. In Hungary, it appeared that governmental controls were less stringent, the chemical plants operating rather independently. We were told, however, that this would not create any serious problems for implementation of the projected Chemical Weapons Convention because such domestic control mechanisms as the treaty might require could be introduced without difficulty.

As regards Yugoslavia, the 'self management' principles under which manufacturing enterprises are now run also mean a certain degree of independence for each chemical firm. A firm may, for example, accumulate up to 49 per cent foreign capital. Nonetheless, the firms are subject to control mechanisms as in other countries, a major recipient of the monitoring data being the trades-union authorities. However, for those plants that have a substantial degree of foreign capitalization, the controls on behalf of the trades-union council are somewhat less stringent than those for other plants.

Developing countries

As to the developing countries, the general picture which we received was of small-scale chemical industry, typically of a multinational character, that tends to concentrate on the production of a few speciality chemicals within a narrow field, and on the formulation and marketing of chemicals manufactured abroad. Acquisition of materials and technology is heavily dependent upon imports. Two major problem-areas, somewhat inter-related, were identified for us: that of enforcing the laws that specified the national regulatory procedures, and that of controlling the activities of multinational corporations. With regard to the former, although national legislation usually provides for an extensive body of regulatory procedures, an insufficiency of technical resources tends to mean that the regulations are inconsistently enforced. Moreover, that domestic law enforcement tends to be dominated more by legalistic considerations than by sufficient appreciation of the technical factors operating in each particular instance. There might be further difficulties, too. For example, the Indian chemical industry is subject to a variety of regulations regarding insecticides, poisons and drugs, as well as legislation on pollution and labour-legislation pertaining to occupational health and safety, these laws being enforced partly at the national level and partly at the state level; but there are often major differences of implementation as between the different states.

This broad picture of the situation in the developing countries prompted the following suggestion from some participants. As compared with their developed counterparts, these countries might well be in a significantly different situation regarding the acceptability of proposed measures of verification under the projected treaty regime. Because their industries operated primarily on the basis of imported technology, know-how and even raw materials, neither economic reasons nor the need to protect proprietary information could, it was suggested, constitute an especially weighty argument against accepting international inspection procedures. Yet, at the same time, precisely because so much of their chemical industry took the form of formulation or of narrow-area speciality-chemical production, non-production verification could be rested with greater confidence upon data-reporting as opposed to inspection techniques. And possibilities existed, some participants suggested further, in procedures whereby an international organ of the treaty kept a record of imported equipments.

Even so, the substantial role of multinational corporations within developing-country chemical industry might still, in the view of some participants, create difficulties for non-production verification. These corporations tend not to be subject to the most stringent routine governmental controls in the developing countries where they operate, nor do their operations abroad tend to be controlled within their parent countries. Possibly, in the aftermath of the Bhopal disaster (wherever the responsibility for it may ultimately be found to lie), there will be significant changes in these respects, perhaps involving more stringent implementation of national regulatory procedures.

Industrial accidents and international safety monitoring

At this point in the proceedings, attention turned to the subject of industrial accidents *per se*. The question was raised of whether benefit might be gained, not least by the chemical industry itself, from the existence of an international body empowered to receive reports of accidents so as to maintain an international surveillance of one aspect of industrial practice. Such a body could also facilitate NPV. Information reported to it might, for example, be allowed to enter the pool of background information with which those other data reported expressly for purposes of NPV in accordance with CWC obligations were assessed at the international level. And if the body were also empowered to monitor industrial safety practices, that too could aid NPV since any production of highly toxic chemicals would require stringent safety measures. There was some discussion, further, of how any such safety monitoring procedures might be defined specifically in order to assist NPV. An example put forward was the screening of certain biomedical parameters among plant employees, such as serum cholinesterase levels. Records of this type, which are maintained as a matter of routine in many chemical production facilities, might anyway be submitted for NPV purposes.

East-West differences in chemical-industry structure

Besides regulatory procedures applicable to chemical processing and product-choice, there are also the controls which exist at the national level on the transportation and distribution of chemicals and their export abroad. National practices in these respects too were described and discussed.

A significant difference between the market and the centrally planned economies was recognised with respect to regulations on what particular chemicals might be produced, marketed and/or exported. In Western and other market-economy countries, it was corporate management within the chemical industry itself that took the initiating decisions on such matters; only thereafter would management decision become subject to the governmental surveillance represented by any applicable regulatory procedures. The point was made, however, that there have in the past been several instances in which chemical firms have volunteered export controls or drawn dubious customers to the attention of their governments. There was thus a self-controlling element within Western chemical industry. But it was remarked by some participants that this element was at its weakest in regard to the large and growing number of small firms that formulate and distribute chemicals produced elsewhere, or otherwise deal in such chemicals. Taken together, these firms constitute a significant part of the chain linking manufactured chemicals to their end-uses. As a chemical proceeds along the chain, its characteristics, description and packaging may change several times; so, too, may its ownership. Even where a particular chemical was in fact subject to governmental regulations, including export controls, the degree to which it was in practice brought under control might suffer from these entrepreneurial activities. The existence of small profit-orientated dealers in chemicals, and even that of small firms marketing speciality chemicals of their own manufacture, was accordingly seen by some participants to be a significant problem for whatever controls might be established as part of the CWC regime. In this connection, there was discussion of the special export controls on named chemicals which several Western governments had instituted following the reports of CW-weapons employment during the Gulf War.

In contrast, the situation regarding production, trade, export and import controls for the centrally planned economies was described as being a much simpler one. The fundamental basis in such economies for any production or marketing activity, including export or import, is the annual plan that is developed both at the central level and at the level of the individual chemical enterprise. Such plans are monitored and approved at the governmental level. Governmental regulations then control implementation of the plans. The GDR Poisons Act was again referred to by way of illustration; the controls embodied in it are performed by independent inspectors, namely representatives of the GDR Ministry of Health. The attention of the conference was also drawn to the number of different regulations and laws that guarantee implementation of governmental control in the GDR chemical industry and to the function of a special governmental commission which controls exports and imports of chemicals.

Techniques applied in national regulatory practices

Discussion then turned to the particular activities that are involved in the monitoring systems currently in operation within the chemical industry. These included the techniques applied by governmental regulatory agencies as well as those used in international regimes such as the WEU system to which the FRG chemical industry is subject (for CW-weapons nonproduction verification) and, in regard to nuclear materials, the EURATOM and IAEA systems. These were controls with which the industry had learned to live, so that, in the view of the Western chemical-industry participants, the procedures involved would meet with no objection if they were to be used within the projected CWC regime as well. The site-inspection techniques discussed included the interviewing of randomly chosen employees, sampling (of air, water, surfaces and end products) and photography--though on this last technique there were some reservations. In short, the majority of the verification proposals that had been put forward in the CD were thought to be tolerable. Nor did these participants feel that site-inspections conducted at very short notice (e.g., 48 hours) as is proposed in the US draft CWC (CD/500) would present any special problems. As to such other techniques as permanent monitoring by inspectors actually in residence at particular facilities or the use of 'black boxes' (permanently emplaced sampling and analysing devices), no substantial objection was voiced against these either, although current black-box technology was not thought to have sufficient long-term consistency yet; but it was doubted whether they would really be needed. As it happens, black-box technology is being increasingly used by the industry itself in the form of continuous area monitoring systems that offer a safeguard in litigation.

The point was also made that, for a particular plant where inspection procedures were to be applied, the advice of the management itself on the best inspection techniques to use should not be neglected: management would have an obvious interest in dispelling any uncertainties in the minds of inspectors, and, from its detailed knowledge of plant operations, would be able to suggest very specific ways in which the requisite proof could be obtained.

The participants representative of Socialist-country chemical industry likewise recognised that use of CWC-relevant techniques by independent inspectors was common practice today within the framework of national monitoring activities in their countries. There was no comment regarding international inspectors.

At this point in the proceedings, after we had heard a succession of interventions by industry participants all indicating the general acceptability of one or another form of proposed control technique, one participant remarked that, if these attitudes were as widespread throughout the world chemical industry as they were among the conference participants, the diplomatical community responsible for negotiating the controls might well stand to be surprised; for a common impression conveyed by such officials was that adverse industry attitudes constituted an obstacle to progress in the negotiations.

Sample-taking, and the concept of negative proof

No participant claimed that the existing structure of control embodied in national regulatory practice was itself sufficient for adequate monitoring of compliance with the projected CWC. Additional measures would be needed. Much attention was paid to possible procedures involving the taking and analysis of samples during on-site inspections of industrial facilities. Here, it became clear from views expressed by industry participants that agreement might be difficult on any such procedures that went substantially beyond those of current regulatory practices. Sampling of plant-throughput in particular, or even of intermediates in transit between production facilities, could reveal commercially sensitive information about production methods and know-how. One way of getting around this problem, some participants suggested, lay in the concept of 'negative proof'.

The objective in inspections based on the concept of negative proof would not be to establish the identity of everything that was being produced or processed at a facility, but would instead be to establish that certain activities were not being conducted, or that certain materials were not being processed or produced - in other words, that certain chemicals were not present in a sample. Such an approach, especially when applied to process sampling, would ensure a low level of intrusiveness into areas of legitimate commercial secrecy. But caution would be needed. For example, analytical techniques are now evolving so rapidly towards ever higher levels of sensitivity that monitoring instrumentation might well give positive findings for chemicals that were in fact present in plant-throughput solely as minute trace contaminants, not as intended reactants or products. It might be necessary, therefore, for the CWC to specify an agreed level of sensitivity for the analytical instrumentation that was to be used in any negative-proof inspection procedures that it prescribed.

Regulatory practices as a basis for CWC verification

There was no dissent from the view expressed by several participants that, all in all, the current monitoring procedures used nationally to implement regulation of chemical industry could provide a good starting point from which to develop the measures needed for international assurance of CW-weapons nonproduction. From the standpoint of the chemical industry, whatever additional monitoring procedures - national or international - that the CWC might provide over and above the existing regulatory regime should have the following characteristics:

(a) The procedures must be, and be seen to be, scientifically sound.

(b) They must have a high probability of yielding conclusive data: they must, in other words, be cost effective.

(c) They must be applicable in a graduated fashion: only in cases where doubts or ambiguities had not been dispelled by application of the simpler procedures should the more sophisticated ones then be used.

(d) They must not endanger the intellectual property of the bodies to which they are applied: they must not, for example, allow know-how or other unpatented information to leak to competitors.

Likewise, the overall CWC verification system should:

(a) be robust enough to withstand changing practices within the chemical industry, including its future technological development;

(b) be simple enough to work reliably;

(c) constitute no obstacle to future scientific and technical progress; and

(d) not interfere with the competitiveness of any individual enterprise to which it is applied.

Whether it would in fact prove possible to devise a CWC regime that could satisfy all these requirements and, at the same time, be effective in implementing a global prohibition of CW armament was, of course, another matter. Crucial would be the specification of precisely which chemicals the treaty was to regulate. This question constituted the second main agendum of the conference.

II.2 Chemicals and chemical production requiring regulation under the projected Chemical Weapons Convention

Applicable criteria

Participants discussed different possibilities for identifying and classifying those chemicals which are to be found in civilian chemical industry and which may require regulation for the purposes of the CWC. The criteria considered included the following:

(a) Purpose criteria, differentiating categories of end-use.

(b) Toxicity criteria, especially the criteria based on acute mammalian lethal toxicity currently used in the CD negotiations.

(c) Military criteria, differentiating categories of potential utility of chemicals as CW agents, including likely military significance if so used.

(d) Proximity criteria, differentiating chemicals according to their proximity as production intermediates to chemicals that might be used as payload of CW munitions.

(f) Quantitative criteria, differentiating chemicals according to, for example, the volume in which they are produced within civilian industry or the number of their different applications there.

There was discussion of these criteria in the context of the 'risk' approach that had been proposed at the CD by the British delegation[6] including its subsequent development in the Swedish proposal of July 1985:[7] an approach towards the differentiation of controls which classifies chemicals according to the risks which they may present for the projected CWC.

The point was made by some participants that the ordering of chemicals which can be achieved by applying criteria such as toxicity, proximity or even military significance would not necessarily correspond to the degree of stringency with which the CWC regime should attempt to control particular chemicals. For that, additional criteria reflecting such factors as civilian utilization or quantities produced for normal industrial purposes would have to be used as well. The point here was, as one participant put it, that controls which might otherwise seem desirable could be rendered impracticable by industrial and political realities; compromises might have to be sought; and the CWC would have to include easily operable amendment procedures. No dissent from this view was expressed. Discussion of possible ways forward jumped to and fro between different aspects of the problem, this itself no doubt reflecting the delicacy of some of the underlying matters; but what we learnt was dominated by a few particular themes.

Desirability versus practicability

On the desirability side of the central issue of desirability *versus* practicability, the interventions made by participants may be summarized as follows:

First, running through all the interventions was a tacit acceptance of the basic approach which the CD is using for classifying chemicals for the purposes of control, which is to say a classification that rests on the 'general purpose criterion' supplemented by toxicity criteria. The essence of this approach is that the proscriptions to be laid down by the CWC should extend to all chemicals except for those covered by permitted-purpose exemptions, and that the controls should be at their most stringent in the case of 'supertoxic lethal' chemicals, these being distinguished by LD50 or LCt50 thresholds from 'other lethal' and 'other harmful' chemicals, to which controls of lesser stringency should, in general, be applied.

Second, there was general acceptance of the principal that, even within the categories delimited by the toxicity criteria, chemicals could vary in the degree to which they ought to be controlled. Chemicals suitable for use as payload for binary munitions and the like, and chemicals that could otherwise serve as precursors of CW agents, were illustrations of this. There might be others also, such as those substances referred to as 'adjuvants': chemicals which, although relatively innocuous in themselves, might serve to enhance the aggressiveness of other chemicals. So there was general acceptance of the need for lists of specific chemicals: lists that would indicate the particular control procedures that were to be applied to each listed chemical.

Third, several participants expressed the view that any such use of lists would have to be complemented by procedures--flexible procedures--for revising the lists as necessary, whether by moving chemicals from one list to another, or by adding new chemicals. The perceived need for such list-revision mechanisms, the operation of which--on pre-agreed principles--might best be left to a Consultative Committee of states parties, was illustrated in several ways. Some participants spoke of the possible emergence of novel types of CW agent, in which regard there was discussion, in particular, of possibilities being opened up by DNA recombinant techniques (see below). One participant remarked that there were several chemicals which had actually been patented as CW agents, but which had not yet figured, even generically, in any of the lists that had been put forward by CD delegations. And some participants also spoke of a danger that lists drawn up in terms of specific chemicals rather than families of chemicals might create loopholes through which ingenious chemists might design new CW agents or CW-agent precursors that would evade the controls in question; the analogy here was to the 'designer drugs' which are currently vexing the US Drug Enforcement Administration.

Fourth, some participants spoke of the desirability of placing such stringent restrictions on certain chemicals, for example certain supertoxic lethal ones, or some of the possible reactants therefor, that their occurrence within the world chemical industry would largely be suppressed. It was generally recognised, however, that outright prohibition of the production or use of any one chemical, or class of chemical, would not be desirable since it would place unwarranted obstacles in the way of scientific research, not least research for medical purposes or antichemical protection. There was general acceptance, then, of the approach--one that offers at least a partial remedy--which the CD has been pursuing in the concept of tightly controlled small-scale permitted-production facilities. Disagreement, however, existed among participants (as is described later) on how this concept should best be applied.

Practicability versus desirability

Turning now to the practicability side of the issue, one set of comments made by participants had to do with control procedures and their feasibility, while another set had to do with the nature of chemical industry, and hence the impact which control procedures might have on industrial practices. The two sets overlap. It is convenient here to report the second set first.

We heard from industry participants about the sort of research that typically underlay decisions to produce new chemical products. A single new pesticide or pharmaceutical product, for example, was now likely to result from research in which maybe 20,000 chemicals had been synthesized and screened. Searches less extensive than this were unlikely to reveal worthwhile new products. This would mean that, if limits were imposed on the areas in which new products might be sought, the chances of finding them, and therefore of benefitting from the new applications or other possibilities which they might open up, would be reduced. Particularly in the pesticide field, the searches can hardly avoid areas of chemistry where lethal or otherwise harmful chemicals abound, including supertoxic ones. The same is true in the pharmaceutical field, for drugs that may be beneficial to sick people may be harmful to healthy people. Therefore, any restriction which the CWC might place on the manufacture or testing of lethal or harmful chemicals could be detrimental to otherwise legitimate and socially valuable industry. Not only would this consideration militate very strongly (as noted above) against any outright ban under the CWC on particular chemicals; its practical implications would also need to be taken most carefully into account in any application of the concept of specially controlled permitted-production facilities; a matter to which we revert later.

For these and other reasons, industry participants also spoke of the dangers of over-long listings of chemicals for control purposes. If such listings included chemicals which were already produced within the civilian chemical industry for legitimate purposes, that production could be hindered. Or there might be other such negative effects: producers of listed chemicals might, for example, decide to abandon production in favour of other products that were not encumbered by controls, thereby adversely affecting end-users dependent upon the existing production. In summary, the view expressed by these participants was that the control lists should be short, and preferably should not contain any chemicals in extensive civil use.

This last point was made not only in relation to chemicals having numerous civilian applications but also in relation to chemicals produced in large volume for a small number of applications. Both categories would also, it was suggested, create problems for the CWC regime from the standpoint of the feasibility of control procedures. In this regard, some participants referred in particular to controls taking the form of data-reporting and statistical accountancy: the larger the scale of production or application of a chemical, the more difficult to operate would such controls become, and therefore the less reliable. For example, the resources needed to operate a materials-accountancy monitoring system which extended to a chemical such as phosphoryl chloride would be enormously greater than for a sytem in which such widely used manufacturing intermediates were not on the control list.

As to control procedures involving site-inspections at industrial facilities, several participants dwelt on the relationship that would exist between the feasibility of such controls and the number of facilities that were to be subjected to them, the latter number itself being related to the number of chemicals listed for such control. A long list of chemicals could imply a large inspectorate, a point which one participant illustrated by reference to the IAEA. The inspectorate of the IAEA, which was not a small organisation, was tasked to apply safeguards in respect of a list of materials that was in fact rather short. However, as discussion on this point developed, the view was expressed that, on the basis of what was known about the actual number of producers of the chemicals specified in the proposed control lists around which consensus within the CD seemed to be beginning to form, the number of facilities that might need to be subjected to an inspection regime might be in the range 50-100 facilities worldwide; comparable, in other words, to the number of facilities subject to IAEA safeguards.

The shape of a possible compromise

All in all, then, we feel that a rather positive conclusion may be reported here, namely that a compromise between desirability and practicability may indeed be attainable, and, moreover, that the compromise need not necessarily be a bad one in the sense of creating either an unworkable regime, or a regime with too many loopholes or one which may hinder beneficial industrial development.

The discussion (as opposed to some of the papers presented) did not, however, proceed into much detail as regards specific chemicals. The comments we heard about individual chemicals and the places which they should occupy on control lists were few, being illustrative more than systematic. However, from the directions taken by the discussion of general principles, we were able to form some impression of the shape which the compromise might eventually take, this being set in our minds by convergences that seemed evident in the views expressed by the industry participants. The salient features of the latter may be summarized as follows:

First, the NPV provisions of the CWC verification regime, be they national or international in their operation, must from the start be as unambiguous as possible. That NPV regime must therefore be simple and robust, relying on 'negative proof' procedures (see above) as far as possible, and it must include easily workable mechanisms for amendment. A regulatory system that is too detailed to begin with would be liable to induce an ineffective regulatory bureaucracy; and a system that is too invasive of legitimate commercial secrecy may fail to secure the fullest cooperation of industrial corporations. If experience shows that the system needs more sophistication in order to afford adequate assurance, then the built-in amendment mechanisms should be used to provide it.

Second, it should neither be the purpose nor the effect of the regime to limit production of chemicals that are currently being produced on a substantial scale for civilian purposes; nor should the regime seek to alter the pattern of the production base for such chemicals by disallowing their production outside specified facilities. This point is discussed further below.

Third, some industrial chemicals and/or activities will nonetheless have to be monitored. In all such cases, the greatest possible precision must be used in defining what is to be controlled. This requirement is bound to operate in tension with the danger that over-precision may create loopholes. But, although for the latter reason the control lists may have to include families of chemical grouped on the basis of rather general criteria, effective implementation at the industry level will nevertheless require that listings of groups of chemicals be transformed into listings of individual chemical species.

In regard to this last problem, it was suggested that the remedy might perhaps be to establish the reporting requirements in terms of families or groups of chemicals but the subsequent monitoring procedures in terms of species-specific lists. For example, if a country or a chemical corporation made a positive declaration of production for a particular group of chemicals, the next step would be for the control organ that had received the declaration to ask for the identity of the particular species within the group that was or were being produced. Guidance on such an approach was available from the current experience of some countries in the promulgation and implementation of export controls on chemicals having utility as CW-agent precursors. We were told how the government of one of these countries had promulgated its Gulf-War-related chemical export controls in terms of families of chemicals, but had then be obliged to make available to its Customs authorities a much longer list of specific chemicals within those families.

Some further details emerged from two specific areas of the discussions just reported which should be recorded here. They concerned DNA recombinant technique, and the concept of specially controlled permitted-production facilities.

Aspects of emergent gene technology

Some participants had expressed concern about the impact which current and foreseeable developments in the application of DNA recombinant technique to industrial processing might have on verification issues. We were told that scientific as well as technological developments in recent years had created a situation in which the entire world now had access to the essential prerequisite of this new biotechnology, namely restriction enzymes; and that high-containment laboratories comparable to the American P3 and even P4 standards were spreading significantly. We were told, further, that modern microbiological culture technology had achieved major successes in the miniaturization of production. This now enabled continuous-flow fermentation for industrial purposes in relatively small devices: production facilities that had once operated batchwise with fermentors several cubic metres in capacity could now use continuous-fermentation technique in reactors a hundred times smaller. Such facilities could, moreover, be converted easily and rapidly from the processing of one organism to another--flexibility which no chemical process plant could match.

The view was therefore expressed that, if there were any products of such biotechnological processing which might be considered to have serious significance for the CWC (toxins, for example), controlling the facilities where such processing might be conducted would very probably be impossible, however much effort was devoted to the task. There was, however, no discussion of the likelihood with which biotechnological-process products might, from the military standpoint, be considered worth weaponizing; nor, if they were to have military attractions, of the scale on which they would need to be produced.

Permitted production within specially controlled small-scale facilities

This matter of DNA recombinant technology also had an obvious bearing on the question of small-scale permitted-production facilities. The discussion of this question took as its point of departure what seems to be an emergent consensus within the CD. Given that an outright ban on production of the most menacing chemicals and their most important production-intermediates would not be possible in view of the legitimate and sometimes essential uses to which any of these chemicals could in theory be put, it would be necessary instead to place a low limit on the aggregate amount that could be produced--one tonne per year per state party has been a long-standing suggestion--and develop tight control procedures whereby states parties could assure themselves that the limit was not being exceeded. To facilitate such control, CD negotiators have introduced the concept of concentrating production within a limited number of facilities worldwide, the locations of the facilities to be declared, and the facilities themselves to be subject to stringent control, including the use of systematic international on-site inspection procedures.

There are two main matters still unresolved at the CD, and it was these which chiefly engaged the conference participants. First, should any production of the chemicals in question--i.e. those chemicals requiring the most stringent degree of control that the CWC can provide--be permitted outside these special facilities, and, if so, how should that production be controlled? For example, was it or was it not sensible to require that the special facilities be used solely for the production of chemicals needed to study and develop antichemical protection, leaving the matter of production for permitted purposes other than protective purposes to some other means of control? Second, and related, should the chemicals that are to be produced, and thereby controlled, within these facilities be defined generically or specifically?

Those participants who expressed themselves maintained that the controls represented by the small-scale facility concept should apply only to listed chemicals, not to all chemicals whose toxicity exceeded the 'supertoxic lethal' threshold and their most important production intermediates (however the latter might be defined). Two considerations were cited in support of this position. First, in the case of pharmaceuticals which happened to surpass the 'supertoxic' threshold in their unformulated state, the existing quality-assurance procedures governing the trade and export of medicinal products might be impossible to reconcile with the requirement that the chemicals in question be produced in facilities also used for production of deadly poisons. Moreover, the requisite scale of production of such pharmaceuticals might exceed the one tonne/year limit. In short, any use of the small-scale facility concept which affected the operations of individual pharmaceutical corporations could have a seriously adverse effect upon them, not least from the standpoint of their international competitiveness. Such are the R & D costs involved in developing new products and satisfying the regulatory requirements for them that all pharmaceutical corporations--whether in the industrialized or the developing countries, East or West--aim at export as well as domestic markets. Second, quite apart from commercial production, it would be unreasonable and probably unworkable to require that pesticide or pharmaceutical firms (for example) conducted a part of their candidate-product screening and development work within the special facilities--i.e. that part of the work which involved the synthesis of R & D quantities of new chemicals which might turn out to be supertoxic. Indeed, and more generally, these participants felt that no laboratory-scale production of chemicals should be encumbered by controls under the CWC. The discussion of what was to be understood, quantitatively, by 'laboratory scale' was inconclusive.

II.3 Possible Implications for the chemical industry

The third agendum of the conference broke itself down into two main fields of discussion: the likely impacts of a future CWC on civilian chemical industry, and the likely impacts of the characteristics of that industry on the overall CWC verification regime. Many of the matters raised had already been touched upon earlier in the conference. We have thought it useful to begin our report of what was said by grouping together those points made by participants which contrasted the Western-country and the Socialist-country attitudes towards CWC verification in its application to civilian chemical industry.

Western attitudes

The Western chemical-industry participants continued to express the view that, for them, most if not all of the procedures and mechanisms currently being discussed within the CD for purposes of CWC compliance verification would be tolerable, since comparable procedures were common practice in the multitude of regulatory processes and governmental controls to which their industry was already subject. They could see no major problems in the controls being applied by international inspectors, provided certain criteria were met, above all as regards the safeguarding of proprietary information and other intellectual property. In this connection we were told that the FRG was able to accept the WEU controls (which have several features in common with schemes currently under discussion at the CD) because the FRG chemical industry was adequately assured that sensitive information obtained by the WEU Armaments Control Agency would not find its way to industrial competitors.

These participants also expressed themselves on the question of whether the chemical industry should be subject both to routine international verification and to on-challenge procedures. Their general view was that routine inspections, provided they were limited in number and extent as by means of the 'random' approach and conducted in accordance with agreed procedures, would be much more convenient for states parties than challenge inspections, as well as being less provocative and less intrusive. These participants appeared to accept the view, however, that routine inspections of civilian chemical industry would be unlikely to afford a high probability of detecting violations of the CWC, though the inspections could certainly create obstacles for any clandestine CW-weapons production within declared facilities. Routine inspections would thus be more in the nature of a compliance-monitoring technique than a compliance-verifying one. Challenge inspections offered the possibility of a safety net.

Socialist-country attitudes

The approach of the Socialist countries as it was expressed during these discussions differed considerably. The Western view that routine inspections on a random basis could be less provocative and less intrusive was not shared by these participants, who took rather the opposite view. International verification procedures in the chemical industry on a routine basis were not seen to be necessary by people in the Socialist countries. They put forward several reasons, in particular reasons such as the complexity of this branch of industry, its international links in the commercial area, the need to protect industrial secrets and sensitive information, and the fact that complete verification of what these procedures would seek to verify would in fact be impossible, given the present state of scientific and technical development. These participants accordingly stressed the view that routine international on-site investigation of chemical industry would not serve the function which their Western counterparts had suggested it would; insistence upon its use might accordingly be taken to indicate an unjustifiable mistrust.

Hence, then, the view that this problem was basically political in character, and therefore resolvable only by negotiation at the CD. But whether its ultimate resolution took the form of an emphasis either on routine national verification with reporting to an international body established by the CWC or on randomly chosen international on-site inspections, the purely technical aspects of the problem did not seem to be limiting factors.

Developing-country attitudes

There were no particular observations made by participants in this area of discussion which could be seen as characteristic of developing-country attitudes. However, it was from the developing-country participants that we heard expression of one of the more fundamental considerations concerning nonproduction verification: if the chemical industry of a country were to produce chemical weapons, would not the initiating action have come from the government itself? How, then, could industry controls--mediated as they would have to be, at least in part, through governmental agencies--inhibit any such production? Would not the application of controls to private industry, then, be pointless? We heard both general responses to this, and ones specific to the situation in developing countries.

As to the former, some participants laid stress on the assurance function of NPV: procedures would be needed whereby states could demonstrate to the world at large, including adversarial or otherwise suspicious states, that their chemical industries were not in fact engaged in CW armament. Complete assurance on this score might be difficult to project, but, with a good NPV system, mistrust could be reduced and confidence built. As to aspects peculiar to developing countries, there was reference again to the activities of multinational and transnational corporations. It was not inconceivable, some participants felt, that such corporations might make use of their facilities in one country in order to satisfy a market for CW weapons existing, not in that host country, but in some other country.

Other considerations: the inspectorate

There was discussion of the kinds of inspectors (national or international) who would be needed for NPV purposes in the chemical industry. As compared with other verification tasks for which inspectors might be needed under the CWC--for example in connection with the destruction of CW weapons or of the facilities that had formerly produced the weapons--rather different qualities would be required, not least because NPV procedures would run indefinitely. Having regard to the areas of confidentiality in which NPV inspectors would be operating, some participants emphasized the personal qualities of integrity and reliability that should be required of the inspectors. As to technical skills, there would be a need for an inspection team to include, as one participant put it, 'educated eye-balls'--i.e., people with several years of plant experience, especially the sort of experience that is gained by trouble-shooters--and, as another participant put it, 'piff', a well developed capacity for sensing something odd. Beyond that, participants laid stress on experience in comparable control duties, but did not see detailed knowledge of the particular technologies involved as necessarily being an essential requirement; the types of inspectors currently employed by local or national authorities in the field of environmental protection might, for example, prove suitable. Recruitment of inspectors from within the industry itself (for example, people from the internal audit teams which some large Western corporations maintain in order to assure themselves of compliance with regulatory requirements) was another possibility discussed by participants, but one that could create difficulties from the standpoint of commercial confidentiality. Indeed, something of a dilemma existed which this consideration illustrated: people who might be thought best qualified for inspection duties by reason of their industrial experience could also be the people most capable of violating intellectual property while discharging those duties. There was general agreement that, if properly conducted, the job of inspectors would be a very hard one, and would need people with a strong sense of vocation.

Participants noted that, as regards international inspectors, experience was available from IAEA safeguards to guide the choice of mechanism whereby the international organ of the CWC and individual states parties could agree on mutually acceptable inspectors.

Among those participants who spoke on the subject, the preference seemed to be for the establishment of a corps of professional inspectors rather than the maintenance of a register or roster of potential inspectors which could be drawn upon as necessary for CWC control purposes.

Further aspects of commercial confidentiality

Commercial confidentiality had been a recurrent topic throughout the conference. The importance of ensuring that the CWC regime did not jeopardise intellectual property had been stressed by participants, especially from the standpoint of protecting such property from leakage to competitors. An instance was referred to in which a particular corporation in the Western world had suffered very substantial damage indeed due to an inadvertent release of data reported in confidence to a regulatory agency. Nor was concern for competitiveness limited to industrial corporations operating within the market economies. Socialist-country participants, too, commented on the importance of safeguarding intellectual property, and during an earlier part of the conference we had heard about some of the characteristics of competitiveness within and between centrally planned economies. We were told, for example, that in one CMEA country, chemical firms protected their individual competitiveness by not sharing their intellectual property with one another. There was such sharing within the GDR, however, but this did not extend to disclosure of proprietary information to chemical firms in other CMEA countries.

One particular form of intellectual property was given much attention during this later stage of the conference, namely the customer lists maintained by chemical manufacturers. Participants recognised that the opening of such lists to a control agency could greatly facilitate NPV controls that relied on materials accountancy, for it would allow controlled chemicals to be tracked through an economy with greater confidence than would otherwise be the case. However, such customer lists are, we were told, guarded most jealously: not only could their disclosure provide valuable market intelligence to competitors, but it could also be detrimental to the competitiveness of the customers themselves, especially when the disclosure extended to rates of shipment of particular chemicals as well as their identity and destination.

One industry participant took the view, apparently shared by others, that, considered in terms of the security of intellectual property, requirements for disclosure of customer lists would be regarded by the industry as a greater threat than any on-site inspections to which the industry might become subject under the CWC. Even so, there were some participants who maintained that the information which industry should be required to declare under the terms of the CWC should include, not only data about producers and production of controlled chemicals, but also data on consumption, consumers and their precise sources of supply as well. An intermediate position voiced by one participant was that both consumption and production data should indeed be disclosed, but information on links between particular producers and particular consumers should not be subject to declaration. Discussion did not, however, penetrate this issue at all deeply.

One proposal which was made was that industrial chemical corporations--East, West and Non-Aligned--should publicly declare their unwillingness to participate in the production of chemicals for CW-weapons purposes or to provide information to other corporations or governments regarding such production.

Wider issues

As the conference approached its conclusion, participants turned their attention to aspects of the various problems they had been discussing which might be tackled within frameworks lying outside or complementary to that of the projected CWC. Two topics in particular were addressed: the notion of CW-weapons-free zones, and CW-weapons proliferation.

We were told about the proposal that had recently been put forward, as a joint initiative by political parties in the two German states (the SED of the GDR, and the SPD of the FRG), for establishing a zone free of chemical weapons in Europe. The approach that had been adopted on the question of verification was explained to us: we learned that the proposed verification machinery would be overseen by a standing consultative committee of the participating states, and would involve a mix of national and international on-site inspections both on a challenge basis and on a routine basis. The governments of Czechoslovakia and the GDR had recently proposed to the FRG that the topic should now enter intergovernmental negotiation.

During the ensuing discussion, several of the problems intrinsic to the free-zone idea were addressed, including the geographical extent of the proposed zone and its verification. An attitude expressed by several participants was that the primary significance of the SED/SPD proposal lay in the political field, and that the initiative which it recommended should be considered as a practical proposal whereby states could manifest the good faith in their protestations of support for CW disarmament. However, other participants took the view that the proposal--even though it had been presented as an initiative that would complement, not substitute for, the CWC negotiations--could detract from efforts at the CD to reach a global ban on CW weapons. They had further objections, too: there would be a geopolitical asymmetry inherent in the proposed zone; and, as regards verification, negotiation would face exactly the same difficulties and obstacles as had impeded the CD--resolvable, therefore, only if there were also to be a breakthrough in Geneva. These and other objections evidently underlay the generally negative governmental reactions to the proposal voiced from the West. It was unfortunate, in the view of some participants, that these reactions might suppress serious attention to the proposal, thereby denying the opportunity which it presented for assessing this particular type of disarmament process in action.

With regard to the problem of the potential proliferation of CW weapons around the world,[8] there was general agreement among all participants that this was indeed a matter for serious concern. All possibilities should therefore be explored for constraining any such spread. It was also generally agreed that such proliferation could not only create serious security problems in certain areas of the world; it could also place major new obstacles in the path of the projected CWC. One view put forward was that the remedy did not lie, however, in the negotiation of a CW nonproliferation treaty; it lay instead in the rapid conclusion by the CD of the projected Chemical Weapons Convention.

Notes and Literature References

1 Brazil. Ministerio da Industria e do Comercio, Conselho de Desenvolvimento Industrial, Secretaria Executiva. Relatorio Anual de Atividades Grupo Setorial III: Industrias Quimicas, Petroquimicas e Farmaceuticas, 1984 (Brasilia-DF, August 1985).

2 Paul Elstner, Karlheinz Lohs and Ursula Stephan (editors), Umgang mit gefährlichen Stoffen: Rechtsvorschriften (Berlin, GDR: Verlag Tribüne, 1982).

3 The Annex to the paper by Dr Carpenter that is reproduced in Volume 2 is an excerpt from a large compendium of forms and other materials illustrating the state and federal controls to which the Monsanto Company, like any other US chemical manufacturer, is subject. The compendium was provided for circulation among the conference participants.

4 This commission was made possible by the John D. and Catherine T. MacArthur Foundation, which, under its International Security Program, is supporting work on chemical and biological weapons arms control at the University of Sussex in a joint project with Harvard University. That support, which also made possible a part of the contributions of one of us (J.P.P.R.) to the conference, is most gratefully acknowledged.

5 United Kingdom. Working paper for the Conference on Disarmament. 'Verification of non-production of chemical weapons: proposals for inspection procedures and data exchange'. CD/575 of 6 March 1985. (Reproduced in Volume 2 as an appendix to the paper by Dr Cooper.)

6 United Kingdom. Working paper for the Conference on Disarmament. 'Verification of non-production of chemical weapons'. CD/514 of 10 July 1984. (Reproduced in Volume 2 as an appendix to the paper by Dr Cooper.)

7 Sweden. Working paper for the Conference on Disarmament. 'A comprehensive approach for elaborating regimes for chemicals in a future chemical weapons convention'. CD/632 of 20 August 1985. (Reproduced in Volume 2 as an appendix to the paper by Dr Lau.)

8 The following press report on this topic was reproduced for the conference at the request of one participant: D. Oberdorfer, 'Chemical arms curbs sought', Washington Post, 9 September 1985, page A5.

THE INDUSTRIAL CONTEXT OF CHEMICAL WARFARE

M L Burstall (Industrial Chemistry Group, University of Surrey, UK)

I Industrial potential for chemical warfare

Chemical warfare (CW) involves the use of noxious chemicals which have to be made on a substantial scale. In principle such CW agents might be produced by the local chemical industry, either in the course of its normal activities--chlorine and hydrogen cyanide are examples--or under contract to the users. Alternatively, they could be imported from elsewhere. Less direct methods might be used. A harmless intermediate produced on a commercial scale for peaceful purposes could be turned into a CW agent at a military establishment; thus has ethylene in the past been converted to mustard gas. Largely self-contained facilities may be brought into being to carry out processes to which there is no civilian parallel, as has been the case with the nerve gases.

Whatever the strategy of supply adopted, however, there must be some degree of interaction with the local industry. At the minimum, some chemicals and plant will be bought even if these are only reagents or laboratory equipment. More generally, the local, often the international, stock of knowledge and skill will have to be drawn upon if the plant is to work and the product is to kill. This knowledge and these skills are to no small extent embodied in the local chemical industry itself. If this is large and sophisticated, then the potential for chemical warfare is large. If the industry is small and backward, it is limited.

This paper will review the current industrial potential for the production of CW agents, and attempt to identify factors which are directly and indirectly related to it. It will consider more briefly the ways in which technical change in the future might alter the nature and distribution of this capacity. As far as possible a world-wide perspective will be adopted, and possibilities of monitoring CW activity emphasized.

II The nature of the chemical industry

The operations of the industry

The chemical industry is a statistical artefact, and definitions of its activities vary considerably. In almost all countries, however, it is taken to include those sectors concerned with the manufacture of plastics, fibres and rubbers, of paints, of fertilizers and pesticides, of soaps, detergents and toiletries, of pharmaceuticals, of dyes and pigments, of photographic chemicals, inks and explosives, and of the host of ancillary materials needed to make and use these final products.[1]

Despite its heterogeneous nature, the various parts of the industry have certain broad features in common. The industry is markedly research-intensive, accounting for 15-20 per cent of industrial R&D spending in most developed countries. In some sectors--notably pesticides and pharmaceuticals--research budgets approach or exceed 10 per cent of sales revenue. Competition through the introduction of new products and processes plays a major role in most parts of the industry, and chemistry is an eminently applicable science. In all countries the chemical industry is a leading employer of qualified scientists and engineers, and an important source of specialized expertise.

The industry is also capital-intensive, and normally absorbs 20 per cent or more of industrial spending in this field. The industry uses continuous-flow processes and fully automated equipment wherever possible. Moreover, there are large inherent economies of scale in chemical production, and for many years the trend has been to use a few large plants rather than many small ones whenever this is feasible. Conversely, the industry employs relatively few people in proportion to its size. Of its workers, a large minority are non-manual, and an unusually large number are qualified in one way or another.

At the same time, no company does everything itself. Many activities are usually carried out by the firms themselves, but many others are bought in. An obvious example of the latter is the supply of raw materials and intermediates, but, of course this practice is widespread everywhere. At least within the OECD countries, all companies bar the smallest do their own research and development, but most put out the design and construction of chemical plant to companies which specialize in this kind of work. Similarly, most laboratory equipment is bought from specialist suppliers. In these ways companies interact with other parts of the chemical industry and with organizations outside the chemical sector.

In the developed countries the industry is subject to a considerable degree of government surveillance. In the West, most of this attention is directed towards the environmental impact of chemical production, which is considerable. The industry has not always been as responsible as it might have been, and has paid the penalty in terms of increased public visibility. At the same time, most parts of it--the pharmaceutical sector is the major exception--are neither subsidized nor controlled economically, and operate in a broadly free market.

The structure of the industry

The world chemical industry is made up of many thousands of firms, but a limited number of large companies exert a dominant influence. These enterprises are all based in the developed countries, most particularly in France, West Germany, Japan, Switzerland, the UK and the USA. Typically they have strong interests in a limited number of product areas to which they remain committed for many years. Equally typically, they are organized on a multinational basis, and have local production affiliates in many parts of the world. In this respect they play an important, if partly unintended, role in the transfer of chemical technology.

Since the Second World War the chemical sector has grown much more rapidly than other parts of manufacturing industry. In most advanced economies it now accounts for 3 - 5 per cent of national income. However, the last decade has seen a pronounced deceleration in growth, and most parts of the industry are approaching economic maturity. Increasingly, the range of chemical products can be divided into two broad classes - specialities, characterized by a large knowledge content, and sold at high prices because of their unique properties, and commodities, familiar compounds made by established technology and sold on price alone.

This suggests that the structure of the world chemical industry is likely to change in the future. As table 1 shows, the industry of today is largely concentrated in the developed market economies. That of the developing countries--and, indeed, that of the command economies--is relatively small in size, derivative in technology and biased towards traditional products. The increasing availability of usable chemical technology will improve their position greatly and lead to a much greater dispersion of chemical production. The capacity for innovation, however, will remain concentrated in western Europe, North America and Japan, since at present only they have the necessary social and economic infrastructure.

Chemical industry and CW capacity

What are the implications of these findings for the projected Chemical Weapons Convention? Before considering, in the next section, the directly relevant sectors of the industry, some general points may be made:

1) Since innovation is important to most parts of the industry, its R&D effort is high. A degree of secrecy about new products and processes is common. Not all information is incorporated in patents. Unpublished know-how is often important.

2) Since companies specialize in certain markets, expertise of particular kinds tends to be concentrated in a relatively narrow range of firms. Because research costs are high, the major multinationals are especially significant in this respect.

3) Since economies of scale are large, the number of suppliers of any given chemical is small compared to output. There may be no more than one or two in a nation, and 15-20 world-wide.

4) Since the industry is generally working in a free market, compounds which do not sell are not made. Conversely, something which sells tends to leave traces if it is made on an appreciable scale.

5) Plant and equipment are normally supplied by specialist contractors, who again are limited in number.

6) Since many parts of the industry are approaching maturity, the range of products and technologies, including CW agents and their precursors, available for exploitation is rising.

7) The capacity for serious innovation in products and processes remains concentrated in a limited number of developed countries. This point is further explored in section IV below.

III Capacity and potential capacity for existing CW agents

The major vesicants

As has been noted,[2] the major CW vesicants today are the sulphur and nitrogen mustards, of which the one first used, bis(2-chloroethyl) sulphide, remains much the most important.

These compounds have no large-scale industrial uses, but are readily made from common starting materials by comparatively simple methods. The main routes employed are these:-

$$C_2H_4 + SCl_2 \text{ or } S_2Cl_2 \text{ -------> } (ClCH_2CH_2)_2S$$

$$(HOCH_2CH_2)_2S + HCl \text{ -------> } (ClCH_2CH_2)_2S$$

$$(HOCH_2CH_2)_3N + SOCl_2 \text{ -------> } (ClCH_2CH_2)_3N$$

The uses, and scale and location of production, of the main potential starting materials are indicated in table 2, from which it is clear that there could be serious problems in monitoring the consumption of ethylene and of hydrochloric acid. Ethylene is perhaps the basic industrial chemical, and annual world output exceeds 50 million tons, the majority of which is captive production used to make plastics and other common chemical products. Hydrogen chloride is similarly produced and consumed on the largest scale.

The chlorinating and sulphochlorinating agents commonly employed, though cheap, readily available and widely used, are less universal in their applications. The chlorides of sulphur go mainly into the fast vulcanization of rubber, while thionyl chloride finds its principal application in the agrochemical industry. Both the number of producers and the scale of production are also more limited than in the case of ethylene. Accordingly, the flows of these compounds should be easier to follow, although it should be noted that they are readily made, meaning that inconspicuous captive production could well escape detection.

Given the probable size of the existing world stocks of bis(2-chloroethyl) sulphide,[2] this problem is perhaps of limited urgency. But a country wishing to enter this field could, however, render itself conspicuous if it began to import, say, ethylene or the chlorides of sulphur while lacking a plastics, rubber or pesticide industry.

Nerve gases

The phosphorus industry

The currently important nerve-gases are all based on organophosphorus compounds. As table 3 shows, phosphorus derivatives are, in tonnage terms, one of the principal products of the world chemical industry. However, the large majority of this output is unrelated in most respects to chemical warfare. It consists of phosphatic fertilizers, phosphoric acid, and inorganic phosphates. Organophosphorus compounds make up only a small part --about 2 per cent by weight--of all phosphorus-containing products, and of this minority only a part is chemically related to CW agents.

There are two industrial approaches to extracting phosphorus in a usable form from the phosphatic rocks in which it occurs. The more widely used are the 'wet' processes, in which treatment with sulphuric acid converts the rock to water-soluble calcium phosphates or to phosphoric acid itself:

1. $[Ca_3(PO_4)_2]_3CaF_2 + 7H_2SO_4 + H_2O \text{ ----> } 3CaH_4(PO_4)_2 + 7CaSO_4 + 2HF$

2. $[Ca_3(PO_4)_2]_3CaF_2 + 10H_2SO_4 + 20H_2O \text{ ----> } 10CaSO_4.2H_2O + 2HF + 6H_3PO_4$

The product obtained depends on the amount of sulphuric acid used. With the lower ratio of acid to rock, as shown in case 1 above, the main product is superphosphate, a soluble calcium hydrogen phosphate which is the most important phosphatic fertilizer. With an excess of acid, as shown in case 2 above, the calcium is precipitated as its sulphate and phosphoric acid is obtained. These reactions are cheap and easy to carry out. Superphosphate is made world-wide and is indeed the principal outlet for sulphuric acid. Outside the USA most phosphoric acid and metal phosphates are also made by the wet route.[3]

The alternative electrothermal process leads directly to elemental phosphorus itself. Phosphatic rock, coke and silica are heated in an electric furnace at high temperatures, the phosphates present being reduced to phosphorus according to the equation:

$$Ca_3(PO_4)_2 + 5C + 3SiO_2 \text{ ----> } 3CaSiO_3 + 5CO + 2P$$

Phosphorus obtained in this way is a much more versatile starting material than phosphoric acid, and most organophosphorus compounds are prepared from it. A large part, however, is burned to form phosphorus pentoxide, from which phosphoric acid and phosphates are readily obtained. From a technical standpoint--particularly the purity of the products--this route is superior to the wet process, but in economic terms the balance depends on the relative costs of sulphuric acid and electric power.

From the CW standpoint, elemental phosphorus is of major importance. All CW agents of the organophosphorus group are ultimately made from it (see below), and as yet there is no established alternative to its use. Known producers and capacity for phosphorus production are shown in table 4. In 1983 world production was probably about 850 000 tons, which represented 65 per cent of world capacity. Nearly one-half of this output came from the USA, where production has fluctuated between 350 000 and

400 000 tons per year since the mid-70s. West European and Soviet output is believed to be roughly equal, capacity utilization in the command economies being particularly low.

Most estimates indicate an appreciable drop in phosphorus production in the past decade.[4] In the case of the United States this has been due to a sharp fall in the demand for high-grade phosphoric acid, which has been offset by a gradual rise in other uses, which in 1983 mounted to about 100,000 tons. The situation in other countries is less clear.

Organophosphorus insecticides

The pesticide sector of the chemical industry is of especial significance in chemical warfare, since an important class of its products is related chemically and physiologically to the nerve gases.

These are the organophosphorus insecticides. Discovered in the late 1930s, but not commercialized until the 1950s, they form one of the principal classes of insecticide. Almost all of them have the general structures:

```
R   O           R   S
  P      or       P
R   X           R   X
```

where the groups R are usually connected to the phosphorus atom through an oxygen or nitrogen atom, and the group X through an oxygen or sulphur atom.[5] The phosphorus atom is most commonly introduced by reaction with phosphorus pentasulphide (compound A); in 1980 64 per cent by weight of US production was based on this starting material.[6] Other reagents include phosphorus trichloride (compound B; basis for 22 per cent of US production), phosphorus thiochloride (C; 7 per cent), phosphorus oxychloride (D; 5 per cent), and phosphonic acid (E; 2 per cent). All these compounds are ultimately derived from elemental phosphorus.

P_2S_5	PCl_3	$PSCl_3$	$POCl_3$	H_3PO_3
(A)	(B)	(C)	(D)	(E)

Some 1500 such compounds have been studied[5] although only about 100 have reached the market.[7] Like the nerve-gases, the organophosphorus insecticides disrupt the central nervous system by inhibiting the enzyme acetylcholinesterase, but unlike the nerve gases, they have been designed to be more deadly to insects than to humans. Their advantages lie in their versatility and their relatively short life. Depending on their structure, organophosphorus insecticides may attack most species, as do malathion and parathion, or only a narrow range, as does schradan. They may act as contact or as systemic poisons. Their life is generally short and, unlike the organochlorine insecticides, they do not accumulate in the food chain. Equally, however, both acute toxicity and persistence can be varied within wide limits. Their main drawback is that, given their mode of action, they are inevitably dangerous to handle, especially when the operators are ignorant or careless.

Consumption of organophosphorus insecticides grew rapidly from 1955 onwards, especially after the disadvantages of DDT and allied compounds became apparent. In terms of active ingredients, a 1983 figure of about 150 000 tons seems likely. At the same time, this sector of the pesticide industry has the characteristics of economic maturity. Although world figures for individual types of product are not available, it is clear that total insecticide consumption has stabilized in the developed countries and future growth, if any, is expected from the Third World. It is also significant that the large majority of organophosphorus insecticides were introduced before 1970, and are out-of-patent or approaching that position.[7] Current research interest in these compounds is low.[8]

Nevertheless, this class of product retains great commercial importance. Moreover, as far as CW is concerned, their manufacture is closely linked to that of the nerve gases through the common intermediates used. The major producers are also important sources of expertise about organic phosphorus chemistry and technology. As would be expected, insecticide production is concentrated in the developed countries. France, West Germany, Switzerland, the UK and the USA have large industries which export much of what they make; that of Japan, though substantial in size, is oriented to the domestic market. Most countries in the Third World rely on imports of finished insecticides or of the active materials from which they are made.

A limited number of large companies dominate this market. Information about total sales of pesticides[9] and about the origins of commercially important products[7] suggests that the following firms have substantial expertise in and productive capacity for organophosphorus insecticides:

United States	American Cyanamid, Dow, Dupont, Monsanto, Stauffer, Union Carbide
France	Rhone Poulenc, Ugine-Kuhlmann
West Germany	BASF, Bayer, Hoechst
Japan	Chugai, Sankyo, Sumitomo
Netherlands	Shell
Switzerland	Ciba-Geigy, Sandoz
United Kingdom	ICI

Of these companies, the technical contribution of Bayer and of Ciba-Geigy is particularly noteworthy. All of them are large multinationals with a lasting commitment to the agrochemical market.

Intermediates for nerve gases

The main routes to the principal families of nerve gases are described elsewhere.[2]

For _tabun_ the starting materials are methanol, ethanol, ammonia and phosphorus oxychloride. The first three compounds are among the commonest of industrial chemicals, and only the last could even approximately be described as a medium-tonnage product. The situation as regards the _sarin/soman/V families_ is more complex, since the more important of these compounds all contain a methyl group directly linked to a phosphorus atom. A limited number of commercial products containing a carbon-phosphorus bond are available, but none which are very closely related to the nerve gases.[10] A variety of practicable methods for forming the methyl-phosphorus bond are nevertheless available.[2] They involve the use of diisopropyl methylphosphonate (compound F below), methylphosphonyl dichloride (G), dimethyl hydrogen phosphite (H), or methyldichlorophosphine (K), all of which are ultimately derived from phosphorus trichloride.

$(i\text{-}C_3H_7O)_2P(CH_3)(O)$	$CH_3P(O)Cl_2$	$(CH_3O)_2P(H)(O)$	CH_3PCl_2
(F)	(G)	(H)	(K)

Data on these compounds and on their precursors are presented in table 3. Compounds G, H and K are commercially available, though probably only on an experimental scale in the cases of the first and the last. The immediate precursors to compound F are made in several countries, as are several analogues. Organic phosphites of this type are used as anti-oxidants and for other specialized purposes; US production in 1980 was about 20 000 tons. Clearly, however, the key intermediates for monitoring purposes are phosphorus trichloride and phosphorus oxychloride. Both are made on a substantial scale and are used to make a wide range of products. Phosphorus trichloride--1983 US production 82 000 tons--is used mainly to make pesticides, lubricating-oil additives, and products for use as dispersants and textile processing aids. Phosphorus oxychloride--1983 US production 27 000 tons--goes into the synthesis of phosphate esters used as hydraulic fluids, plasticizers, and surfactants. World production of both these compounds appears to have been static in recent years.

Other agents

The other possible agents or intermediates may be divided into two groups from the industrial standpoint. Some, like hydrogen cyanide and phosgene, are made on a large scale and have substantial civilian uses. Others, like 3-quinuclidinol, although commercially available, are made on only an experimental scale, and have limited uses. Some data relating to these compounds are shown in table 5.

IV Possibilities for the future

Chemical warfare seems to be a mature industry.[2] The principal agents are middle-aged or elderly; indeed, most of them were discovered before 1950. Practicable technologies for making them have been developed and used on a full scale. Large stockpiles have been built up, but demand is essentially low, because the use of CW agents is generally seen as ineffective or unacceptable.

Could this situation change? If so, from where would the necessary expertise and industrial capacity come?

At the present time the most likely development is the use of toxins derived from natural sources. A number of these are already known--ricin is an example--and many more could no doubt be identified if the demand was there. The next stage would be to identify the active species or molecular sites and to use this knowledge to produce improved synthetic or semisynthetic analogues. Alternatively, the large and rapidly increasing stock of biological knowledge could be used to develop such products *ab initio*.

Such courses of action would require a major effort. The potential agents already known are of complex structure and are difficult to collect and isolate. Natural sources of supply would be inadequate for large-scale use. New methods of production would become necessary: biotechnology offers the most attractive possibilities. Military problems connected with the use of the agents as weapons would have to be solved, as no doubt they would be. The development of synthetic toxins would be yet more complex.

The skills needed for this type of work would be quite different from those hitherto used in the CW field. The traditional CW agents are relatively simple in nature and, at least in principle, quite easy to make. The development of practicable synthetic routes was a matter of straightforward experimental chemistry--'dangerous but dull' as a participant described it to the writer--and the construction of the full-scale plants required chemical engineering skills of an entirely conventional type.

In contrast, the development of toxin-based agents and their manufacture on a large scale would need a massive input from scientists and engineers whose orientation was towards biology rather than towards chemistry. The fields of knowledge to be drawn on would be biochemistry and physiology in the development of the materials, and biotechnology in their manufacture. In the latter stage two special skills would be required: genetic manipulation to yield micro-organisms capable of producing the agents, and process biochemistry to culture them on a large scale.

At least a part of the necessary reserves of knowledge and skill are already available. They are to be found in the major centres of pure research and in the pharmaceutical sector of the chemical industry. Drug companies are always keenly interested in new products, and a consensus has emerged that the best way to find them is through the insights gained from the biochemical and biological sciences. Equally important, biotechnology has emerged in the past few years as routes to materials which are otherwise difficult or impossible to obtain.

The research-based pharmaceutical industry is dominated by a group of some 50 multinational companies based in Western Europe, North America and Japan. Those of West Germany, Switzerland, the UK and, especially, the USA are of predominant importance, and account for the majority of significant discoveries. Japanese firms are rapidly improving their position, however, and are particularly strong in process biotechnology. Expertise in genetic manipulation is shared between the major transnationals and a number of small research-only companies specializing in this area. The latter are mainly located in the USA and Western Europe, and exist on a symbiotic relationship with their larger customers.[11]

It therefore seems probable that only the United States and perhaps certain West European countries are definitely in a position to develop and produce novel CW agents of this type. The capacity of the USSR to do so is unknown; current Soviet standing in the international biological-sciences community and the international reputation of the Soviet pharmaceutical industry[12] may or may not be pertinent. The countries of the Third World generally lack the necessary scientific and technological infrastructure; India and Argentina may be exceptions to this generalization.

Other versions of the future are possible. Toxins may not prove to be the preferred CW agents of the future. Other novel but simpler, more readily produced substances may prevail. Yet it is highly likely that whatever substances are used, the biological sciences will play a much larger part in their discovery and development than in the past. Understanding will be more important and serendipity less important than in the period when most of the CW agents of the present day were introduced. If this is so, then the analysis put forward in the present section stands, regardless of which path is followed.

Table 1 Distribution of chemical output, 1980

	USA	WESTERN EUROPE	JAPAN	COMMAND ECONOMIES	THIRD WORLD	WORLD
Population (M)	228	337	117	1491	2275	4500
Per Capita GDP ($)	11350	8890	8880	n/a	670	
CHEMICAL PRODUCTION:						
Sulphuric acid (M tons)	39	24	7	39	18	133
N Fertilizers (M tons N)	12	11	1	17	20	63
Ethylene (M tons)	14	11	4	4	3	35
Plastics and resins (M tons)	10	19	6	7	2	45
Pharmaceuticals (000M $)	18	25	12	14	10	80

Sources: various

Table 2 Precursors of vesicant CW agents, 1983

PRECURSOR	PRODUCTION (000 TONS)		NUMBER OF PRODUCERS					MAIN END-USES
	USA	WORLD	USA	WESTERN EUROPE	JAPAN	COMMAND ECONOMIES	THIRD WORLD	
C_2H_4	14000	35000	19	35	12	?	8	Plastics
HCl	2500	9500	28	<--------	--------	212--------	-------->	Chlorine
SCl_2	10		2	1	2			Rubber vulcanisation
S_2Cl_2	45		2	7	2			Rubber vulcanisation
$SOCl_2$	25		2	10	4			Agrochemicals
$S(CH_2CH_2OH)_2$	<--------	--------	--------	no supplier	listed--------	--------	-------->	

Sources: references 1, 12-14, other.

Table 3 Industrial phosphorus compounds, 1983

COMPOUND	PRODUCTION (000 TONS)		NUMBER OF PRODUCERS					MAIN END-USES
	USA	WORLD	USA	WESTERN EUROPE	JAPAN	COMMAND ECONOMIES	THIRD WORLD	
Inorg. phosphates	9500	36000	<---------------Several Hundred----------------->					Fertiliser, animal foods
Phosphorus	365	850	7	6	2	4	2	See text
PCl_3 (Compound B)	82	200	7	14	6	2	4	See text
$POCl_3$ (Compound D)	27	50 ?	7	14	6	2	4	See text
$MePCl_2$ (Compound K)			2					Experimental amounts only?
$(MeO)_2POH$ (Cpd H)			2	4	1			See text and reference (6); main uses are in flame-retardants, lubricating-oil additives
$(EtO)_2POH$			2	7	1			
$(PrO)_2POH$ (inc. Cpd F)			1	2	1			
Trialkyl phosphites			3	6	1			
$MeP(O)Cl_2$ (Cpd G)			4	1				Experimental amounts only?
$MeP(O)(OH)_2$			1	2				Textile processing
Malathion			6	14	5	3	3	Major insecticide

Sources: references 3, 6, 12-14.

Table 4 Estimated world capacity for elemental phosphorus

COUNTRY	OPERATING COMPANY	LOCATION	ANNUAL CAPACITY (000) TONS)
Canada	Erco Industries	Varennes, Que.	20
		Long Harbour, Newfoundland	62
FRG	Hoechst	Knapsack bei Köln	80
France	ELF	Pierrefitte	11
		Epierre	9
GDR	VEB Stickstoffwerke	Piesteritz	18
India	Star Chemicals	Tana-Belapur	3
	Excel	Bhavnagar	5
Italy	Montedison	Crotone	16
Japan	Nippon Kaguku	Koriyama	15
	Rin Kagaku	Toyama	11
Netherlands	Hoechst	Vlissingen	99
South Africa	Chemfos	Kookfontein	6
USSR	Government	Chimkut	200
		Dzambul	195
		Togliatti-Kuzbyshev	40
USA	Monsanto	Columbia, Tenn.	109
		Soda Springs, IA.	100
	Stauffer	Mt. Pleasant, Tenn.	40
		Silver Bow, Mont.	35
		Tarpon Springs, Fla.	21
	FMC Corp.	Pocatello, IA.	130
	Hooker	Columbia, Tenn.	51
	Electro-Phos	Pierce, Fla.	21
	Mobil	Nichols, Fla.	5
	TVA	Wilson Dam, Ala. (inactive)	32
WORLD TOTAL			**1350**

Source: Reference 3

Table 5 Miscellaneous CW agents and precursors

COMPOUND	PRODUCTION (000 TONS)		NUMBER OF PRODUCERS					MAIN END-USES
	USA	WORLD	USA	WESTERN EUROPE	JAPAN	COMMAND ECONOMIES	THIRD WORLD	
Hydrogen cyanide	180	500	5	5	7			Plastics, herbicides
Phosgene	600							Plastics
Cyanogen chloride								
3-Quinuclidinol		*	2	2				Reagent
1-Methyl-3-piperidinol		*	1	3				
1-Methyl-4-piperidinol		*	2	2				
Benzilic acid			1	11	1			

* Probably in laboratory-scale amounts only.

Source: Reference 14, other.

NOTES AND REFERENCES

1 For general accounts of the world chemical industry see B G Reuben and M L Burstall, The Chemical Economy, London: Longman, 1973; and C A Heaton (editor), An Introduction to Industrial Chemistry, Glasgow: Leonard Hill, 1984, Vol.1.

2 J P Perry Robinson, Chemical-warfare weapons and relevant aspects of their manufacture, background paper CIPCWC/BP.1 DRAFT, 20 May 1985.

3 J R Van Wazer, in Kirk-Othmer, Encyclopedia of Chemical Technology, 3rd edition, 1980, Vol.17, pp.473-490.

4 This seems probable from national data on phosphorus production. In the USA output fell from 550 000 tons in 1968 to 365 000 tons in 1983 due mainly to a decline in the demand for high-quality phosphates used in detergent powders (Van Wazer, op cit).

5 C Fest and K-J Schmidt in K H Buchel (editor), The Chemistry of Pesticides, New York: Wiley, 1983, pp.48-124.

6 I A Boenig, M M Crutchfield and C W Heitson, in Kirk-Othmer, op cit, Vol.17, pp.491-539.

7 C R Worthing and S B Walker, The Pesticide Manual, London: British Crop Protection Council, 1983, 7th edition.

8 Information from a leading independent centre of agrochemical research. Most innovative effort now centres on the synthetic pyrethroids, which are chemically unrelated to the organophosphorus compounds.

9 Predicasts, World Pesticides, New York: Predicasts, 1982.

10 They include a number of other agrochemicals, mainly in the herbicide and growth-regulating groups (cf. reference 7 above). They are all of the alkyl or arylphosphonate structure.

11 See M L Burstall with Ian Senior, The Community's Pharmaceutical Industry - Evolution of Concentration, Competition and Competitive Strength, Brussels: European Commission, 1985, chapter 5.

12 US Department of Commerce, Bureau of the Census. Current Industrial Reports - Inorganic Chemicals, Washington DC, 1983.

13 US International Trade Commission: Synthetic Organic Chemicals, Washington DC 1984.

14 Chemical Information Services, Directory of World Chemical Producers, 1985/6 Edition, Oceanside, NY.

THE CHEMICAL INDUSTRY AND CHEMICAL-WARFARE DISARMAMENT: CATEGORIZING CHEMICALS FOR THE PURPOSES OF THE PROJECTED CHEMICAL WEAPONS CONVENTION

J P Perry Robinson

I Introduction

In August 1985, the governments participating in the 40-nation Conference on Disarmament (the CD) registered a consensus on how 'permitted activities' involving supertoxic lethal chemicals and certain precursors of them were to be controlled under the projected Chemical Weapons Convention (CWC).[1] This agreement was heartening to outside observers of the CWC talks, being a relief from the previous sixteen months of superpower discord and acrimony. For there was real substance to it: the likelihood is that, when the CWC is finally negotiated and enters into force, the overall impact of its regime upon national chemical industries will prove to have been shaped by the August agreement.

The agreement is founded on the idea of correlating categories of chemical grouped according to their importance for chemical warfare (CW) with categories of regulation grouped according to their stringency of control. This is not a new idea, but its formulation in terms acceptable to East and West, North and South, looks like a breakthrough, far from complete though the agreement still is. The main implications for the chemical industry seem to lie in three of its aspects: (a) the control regime will extend to chemicals that are not themselves sufficiently toxic for use as CW agents; (b) one of the control categories will require states parties to prohibit both production and use by their industries of a category that includes such chemicals, save in quantities less than a low limit (one tonne per year); and (c) governments will be required to collect detailed production/use data on a range of industrial chemicals and make periodic international disclosures of quantitative information derived from those data, subject to verificatory procedures run by an international body.

Exactly which chemicals are to fall within which categories, and what the control regimes for each category are to be, will inevitably take much time and negotiation to resolve. At stake are industrial, and hence national, interests on the one hand and, on the other, the effectiveness and viability of the CWC regime. The ideal would be if both regime-effectiveness and legitimate industrial interests could ultimately be satisfied without detriment to either. But the sheer breadth and diversity of the industrial base from which CW capability may be derived militates against this. The requisite compromises will be reached by the traditional processes of politics--bargaining inside, outside and between governments. One may nevertheless express the hope that the compromises will be guided by disinterested analyses of how the control requirements of the CWC regime should best be defined. The analysis presented in the present paper[2] is intended as a contribution to this end.

The purpose of the paper is, accordingly, to identify the chemicals that would be of special concern within a CWC regime, and then to rank them into an order reflecting their relative significance for the regime. This ranking can then serve to focus discussion of the nature and differentiation of the requisite controls.

A logic for those controls is set out in a monograph which the present author wrote in summer 1984, and which SIPRI has just published.[3] It was developed by asking this set of questions: What should be the precise purpose of the CWC? What must be the nature of its provisions in order to allow states to judge themselves better off inside the CWC regime than outside it? What could be the mechanisms that the CWC regime might best provide in order to allow states to assure themselves of the continuing value of their participation in the regime? The answers that were derived were then used to construct a scheme for the scope and verification provisions of the CWC whereby essential elements could be differentiated from inessential ones. The key design principle was that the verification and associated provisions should be able to yield sufficient warning-time for states parties to adopt countermeasures in the improbable event of adversary noncompliance.

Central to the analysis in the mongraph was the matter of the military utility of chemical weaponry in all its varied forms. High-utility weapons have greater implications for a country's national security than do low-utility weapons. The high the utility, the greater must be the assurance which a state can derive from the CWC regime that it will be better off inside the regime than outside it. If the regime embodies controls sufficient to provide this assurance, then pursuing a policy of international collaboration by participating in the regime could make at least as much sense as a policy of national self-reliance by, for example, maintaining a deterrent CW capability.

This logic is also used in the present paper to derive the rank-ordering of chemicals that is sought. The paper rests on the following:

(a) A literature search, seeking to identify all those toxicants that have in the past been used or stockpiled as CW agents. An attempt was also make--necessarily an incomplete one--to identify other toxicants having serious candidacy today as CW agents. Only open-source publications and documentation were used. How severe a limitation that may be is for others to judge.

(b) Application of principles likely to direct weapons-choice so as to identify which of the toxicants disclosed by (a) could be regarded as the more important ones. These are the chemicals labelled below as major 'threat agents'--'threat', that is, to the projected CWC regime.

(c) A second literature search, seeking to identify the plant-scale and binary-munition production routes that have in the past been used or developed for the major threat agents. Then, a compilation from that literature of the principal chemical reactants involved as production intermediates.

(d) A broadening of the list of precursors from (c) to accommodate:
   i plausible alternative production routes,
   ii homologues of the end-product CW agents, and
   iii those industrial chemicals that today could still have some utility as CW agents.

(e) Application to this list of two ranking principles:
   i the same concepts as were used in (b) so as to rank the listed chemicals according to the military significance of the CW agents they could provide; and
   ii in the case of the listed precursors, the number of production steps away from CW agent.

The outcome is presented (in table 9) as a set of five groups of chemicals, the groups being ranked in descending order of what had thus been judged their relative significance: the degree of attention which they warranted in the design of the CWC control regime. The paper closes with a discussion of some aspects of the ranking that bear upon the feasibility of actually applying the controls which it suggests may be needed. A parallel is drawn with the chemical export controls that some governments have recently introduced for the purpose of inhibiting international diffusion of CW-weapon technology.

The paper begins with sections providing background information on the general nature of chemical warfare and on some relevant technical matters.

II The main forms of chemical warfare[4a]

CW is the use of toxicants as weapons of war. The toxicants (or 'CW agents') may be loaded--usually as liquids, but sometimes as solids--into munitions that can be delivered to their targets by general-purpose weapons, such as artillery, missiles or ground-attack aircraft. The toxicant may be disseminated from the munition by means of an explosive or pyrotechnic charge that shatters or distills the payload, by spraying through a suitable nozzle, or by simple rupture of a container. More specialized methods have also been developed, and occasionally used.

In the past, CW operations have fallen into two main categories, direct and indirect. In the direct form, the toxicants have been used to damage the enemy by direct bodily harm to his forces and sometimes non-combatants also. In the indirect form, the primary targets have been the enemy's sustenance: food-plant cultivations, natural cover and economic crops or livestock. In terms of total tonnages of toxicant actually used, the history of CW over the past seventy years has divided itself almost equally between these two categories.[4b]

The feature common to both categories is that the weapons used are 'biospecific'. In contrast to most other types of weapon, they can damage only biochemical processes that support life, whether that be the life of people, plants or animals. They cannot damage materiel, which is to say such typical military targets as the enemy's weapons, installations, fortifications, bridges and the like. At root, therefore, CW weapons are special-purpose weapons, militarily useful only against a relatively narrow range of the targets valued in war-fighting doctrines. Rarity, not frequency, of use has been a prominent characteristic of CW.

In the forms in which they have been most used or stockpiled, the weapons for the indirect category of CW have exploited toxicants maximized for toxicity towards plant life: anti-wheat and anti-rice weapons, for example, or defoliants. It was in the R&D programmes which investigated weapons of these types during the onset of World War II that such herbicides as the phenoxyacetates (24-D, 245-T and the like) originated and then grew into commercial prominence.[5] Since then, civilian agrochemical R&D has, especially during the past decade, provided several herbicides of even greater activity than those that have been used for CW. This paper will not, however, address herbicidal CW weapons any further.

As to the direct category, namely antipersonnel CW weapons, these have exploited two broad classes of toxicant, <u>incapacitating</u> and <u>lethal</u>. It was with the first--in the form of what are nowadays called 'tear gases'--that the CW operations of World War I began. Incapacitating CW agents are ones for which the most likely consequence of exposure under field conditions--which is to say, the most likely consequence of exposure to the dosages which the weapons disseminating the toxicant are likely to impose upon their target-populations--are effects of a debilitating kind from which most victims will eventually recover. Actual fatalities may be low, and will not be the principal objective for which the users have selected the weapon. That objective might be no more than harassment, as in the use of tear gases or other such agents of great irritant toxicity. The harassment might serve as an adjunct to conventional firepower, or to facilitate the manoeuvre of ground troops against defended positions. Or, rather differently, the objective might be enemy casualties, as with the lethal agents, but casualties that would be predominantly non-fatal: the military benefits that stem from attrition of enemy forces might then be augmented by the morale and logistical penalties which large numbers of 'walking wounded' would impose on the enemy. Casualty incapacitants, in contrast to harassing incapacitants, would be chosen for the longevity of their disabling effects. Their toxicity would be one that, at field dosages, threatened motor or cognitive functions rather than vital functions, and which would disrupt such functions for a period much longer than that of exposure to the agent. Their action on the body might be systemic, as with the so-called 'psychochemical' CW agents inspired during the late 1940s by mescal, cannabis and LSD. Or their principal action might, like that of lachrymatory or sternutatory harassing incapacitants, be peripheral rather than central, as with the blinding and skin burning agents.

The distinction between lethal and incapacitating CW agents is not, however, as readily definable as the foregoing may imply. Only very crudely can it be defined in terms solely of properties intrinsic to individual toxicants. Almost all substances, not least the tear gases, can kill if the dose taken into the body from the applied dosage is high enough. And if the absorbed dose is sufficiently low, even the most deadly poison will fail to kill. Thus, so-called 'lethal' agents could, in fact, end up being used for incapacitating purposes--harassing or casualty. Of the main lethal agents used during World War I, the casualties from phosgene and diphosgene seem to have been around 16 percent fatal, while those from mustard gas--twice as deadly but far less volatile--were only 2 percent fatal; for conventional weapons, the overall case-fatality rate was 20-30 percent.

The physical properties of the toxicant used will, in part, determine the size of the dose experienced from a CW weapon following inhalation of the toxicant or its absorption through the skin. So, of course, will the number of weapons used against the target. Propensity to evaporate--called here 'volatility'--is a key property. If the volatility of the agent is low, the airborne dosage set up by the weapon--the degree to which the air around the target population is polluted and therefore threatening--will also be low, unless the weapons used have been specifically designed to aerosolize the agent. Low volatility, meaning a slow rate of evaporation, will, however, favour high contact dosages, for provided the agent is chemically stable in its new environment, it will persist for longer on surfaces with which the target population may come into contact; and if the agent possesses sufficient cutaneous or percutaneous toxicity, the contact dosage (the size of which will be determined by the physical properties governing 'pick up' of the agent from contact with contaminated surfaces) may be great enough to be significantly threatening.

The size of the actual dose taken up from the applied airborne or contact dosage will be determined by the protectedness of the people so threatened. The skin over some parts of the body will more of a barrier than that over other parts against the absorption of an effective dose from a contact dosage; clothing will diminish the absorbable dose to an extent dependent on certain physical and physicochemical properties of the threat agent; and special protective clothing is available for diminishing it further still, even to the point of affording complete protection. Likewise, respirators (gas masks) can provide varying degrees of diminution in the respiratory or ocular dose taken up from an airborne dosage. The more cumbersome forms of respirator and protective clothing may, however, themselves degrade their wearers' abilities to perform military tasks, in which case CW attack might achieve militarily significant effects regardless of the toxicity of the CW agent used.

III The range of potential CW agents

These technical considerations have been recounted in order to convey an idea of just how wide in their variety the chemicals may be that could find application in CW weapons. It is apparent that a whole range of different military purposes could be served by chemicals providing one or another combination of toxic effects, toxicity, vapour pressure, surface tension, viscosity, stability towards environmental factors, detectability, physical state at ambient temperatures, and others. Some of these are properties defining threshold criteria beyond which a chemical has little candidacy as a CW agent; and attempts have occasionally been made[6] to develop general rules that would allow the candidacy of any given chemical to be ranked in terms of an aggregate of its characteristics as determined for these properties. Some are properties defining special requirements in the mechanism of the munition that would be needed to disseminate the agent, or in the manner in which it should be used against its target. Some amount to variables whereby the designers of CW weapons may tailor target effects to particular military missions. For antipersonnel CW purposes, then, there are probably thousands of different chemicals that might in theory serve.

Actual practice this century, however, points to a different state of affairs. In table 1 there is given a reasonably comprehensive list of those chemicals that have actually been used or stockpiled as CW agents.[4b] Table 2 lists chemicals which, although not (so far as is publicly known) actually put into quantity production for CW purposes, have nonetheless either been approved for weaponization or seriously considered. Also shown are several other toxicants which have been publicly stated by officials in one country or another to exist in the arsenals of potential enemies. Although there are no doubt several omissions in these lists, the number of chemicals shown falls far short of thousands. In table 1, a good many chemicals are there simply because World War I battlefields served as the field-test venue which subsequently eliminated their candidacy as worthwhile CW agents.

The presence in table 1 of chemicals which today find heavy application in the civilian chemical industry should be noted. This circumstance of ready accessibility could, in itself, mean that these chemicals must be judged significant threat agents. Phosgene, hydrogen cyanide and cyanogen chloride are the principal ones. By the same token, other toxic chemicals which are in wide civil use, but which are not normally discussed as CW agents, might also need to be considered as threat agents. This is a matter which the Bhopal disaster, not least, suggests should not be overlooked.

It is conceivable, certainly, that there are CW agents in present stockpiles so secret that no inkling of them exists in the open domain. Yet the past history of CW preparations militates against this possibility. At least in the case of the Western countries, all the most closely guarded CW-weapons secrets of the past seem to have surfaced within a very few years, or were to be found in open sources anyway should the scrutiny have been close enough. The German 'nerve gases' of World War II may perhaps be held up as a warning against complacency on this score. However, with hindsight, it can now be seen that the surprise engendered among Allied investigators by their discovery of the German work on tabun, sarin, soman, etc. in 1945 was a reflection, not only of the security which the German Reich was able to maintain around that work, but also of rectifiable deficiencies in Allied CW intelligence.

There is no great sense in having CW weapons unless they are clearly superior to regular weapons (i.e., conventional weapons) in serving the purposes for which they are best suited; and even then there is no point in having them unless those purposes are ones that are valuable under prevailing military strategy and doctrine. The basic criterion, then, for candidate CW agents--for toxicants under consideration for weaponization--is that they should be able to furnish weapons that are competitive with modern conventional weapons in a significant number of valued combat roles. Such as they are, the militarily attractive features of CW weaponry reside in area-effectiveness, search-out capacity against target populations physically protected against conventional weapons, ability to deny terrain --including, for example, maintenance and staging areas--by casualty-threatening contamination of it, and special impacts upon morale. But as the technology of conventional weaponry has advanced, so too have these features of CW weaponry been losing their singularity. New, more precise, target-acquisition techniques diminish need for area weapons; and high-fragmenting and cluster munitions, fuel-air explosives and flame weapons, artillery-scatterable minelets and the new airfield-attack weapons may all afford a greater economy of force than CW weapons. Simple cost-effectiveness considerations have thus come to reduce the attractions of CW weapons for the execution of those military missions, such as attrition, flank-protection or particular types of interdiction, for which they might once have seemed better suited than conventional weapons. Other factors militating against CW stem from the general requirements set by tactical doctrine for the target-effects of weapons, such as requirements for rapidity and predictability of effect. If such requirements cannot be met, CW-use concepts can be integrated only with difficulty into conventional doctrine. The benefits that may nonetheless be available from CW would then be purchasable only at significant opportunity cost to conventional capabilities.

Thus it is that the great majority of CW agents that once figured in arsenals no longer have a logical place in modern arsenals. Toxicants such as phosgene, diphosgene, hydrogen cyanide or cyanogen chloride might conceivably still find military application, but their status as threat agents is no longer what it was during World War II. Conventional-warfare technology and doctrine have moved on. Under today's circumstances, only two families of antipersonnel CW agent can seriously be considered to have a real security significance: the nerve gases and some of the blister gases.

IV The principal threat agents

The nerve gases, which date back to the discovery of tabun at the end of 1936 during industrial pesticide research in Germany, are pre-eminent by virtue of a combination of three properties. The first is the nature of their toxic action, which is such as to kill or disable a person within a very short time of an effective dose being absorbed into the body. A sufficiently large airborne dosage (which may not be detectable by unaided human senses until too late) can kill within less than a minute of inhalation. The second property is ability of a large enough contact-dosage of liquid nerve gas to penetrate through the barrier layers of the skin and then to reach vulnerable sites in the nervous or neuromuscular systems in amounts sufficiently undiminished by enzyme or other biochemical attack to remain effective. By virtue of this property, nerve gases are 'dual threat' agents, capable of creating both an airborne hazard and a contact hazard. The third property is their great toxicity, which is to say the smallness of their effective doses. In comparison with the older lethal agents, the number of munitions needed to attack a given target is thereby much reduced. Many poisons exist which can match, or even surpass, the nerve gases in one or another of these three properties; but none appear to exist that can do so for all three and, at the same time, be stable and amenable to quantity production. It is from their combination of these features that the nerve gases derive such competitiveness as they have with conventional weapons under modern mechanized-warfare conditions.

Chemically, the nerve gases are P(V)-acid esters that are capable of inhibiting acetylcholinesterase enzymes. According to an authoritative US source,[7] at least six families exist that have CW potential, each family having numerous members. General formulae are shown in table 3. The best known nerve gases are from three of these families: tabun (I: R = Me, R' = Et), sarin (II: R = Me, R' = ^{i}Pr), soman (II: R = Me, R' = 1,2,2-trimethylpropyl) and agent VX (VI: R = Me, R' = Et, R" = ^{i}Pr). What are commonly referred to as the 'G agents' are nerve gases in families I and II. The 'V agents' are nerve gases in family VI.

It will be seen from table 3 that, in theory, many hundreds of different chemicals exist that could be called 'nerve gases'. Very probably a high proportion of them have actually been synthesized and studied for their candidacy as CW agents. The number of serious candidates, however, is probably no more than a few dozen, being limited chiefly by the requirements for optimizing the properties enumerated above, not least considerations of accessibility to large scale manufacture (itself a function of the chemical economy within which the manufacture would take place). There may, in addition, be special requirements. For example, some nerve gases give rise to a poisoning that is refractory to treatment with oxime-family antidotes (reactivators of inhibited acetylcholinesterase). If this were seen to be a militarily valuable characteristic, it would mean that those structures in which the O-substituent could yield a relatively stable carbonium ion would be favoured. Soman is especially refractory; sarin and VX are not.

Alongside the nerve gases as major constituents of today's CW arsenals are various blister gases or 'vesicants'. These retain some of the importance acquired by their progenitor--mustard gas--when it was first used in 1917, and the Iraq-Iran war has provided evidence of continuing military utility today. That utility is narrower than that of the nerve gases since the most amenable vesicants are all of low volatility, and are slow-acting. But within these limitations, certain of the vesicants (above all mustard gas) have one aggressive characteristic which the nerve gases lack: in warm weather, simple munitions can readily establish a casualty-threatening airborne dosage to which even people wearing respirators may be extremely vulnerable if they are compelled to remain in the contaminated area.

There are two main classes of vesicant: the mustard-gas families, and the dichloroarsines. The latter, such as lewisite and phenyldichloroarsine, have faded greatly in importance. The mustards fall into two main families: thioethers and amines containing at least two 2-chloroethyl groups--the sulphur-mustards and nitrogen-mustards, respectively. The sulphur-mustards are the more important, and, of them, it is the original one--*bis*(2-chloroethyl) sulphide--that remains dominant. The total quantity of it that is stockpiled around the world today is probably on the order of a hundred thousand tons. Formulations of it that contain around 40 percent of the higher mustard *bis*(2-chloroethylthioethyl) ether (agent T) have been, and are still, heavily stockpiled as fill for spray-type munitions or for cool-weather operations.

V Quantities of CW agent having military significance

The absorbed dose of nerve gas having a high probability of killing a adult person can be estimated, from animal experiments, to be about a milligram, some nerve gases more, some less. In theory, then, a million people could die from a kilogram of nerve gas. With mustard gas, a million theoretical lethal doses could, likewise, amount to about 15 kilograms. Yet, according to a US statistical study,[8] it took on average more than 25 kg of mustard gas to produce a single casualty during World War I; and the great majority of mustard casualties then were non-fatal. There is nothing peculiar to poison in this. For small-arms fire, expenditure seems to have been in the region of 5000 bullets per World War I casualty; nowadays it would probably be far higher.

As has been seen, the dosage-field established by functioning CW munitions is determined by a wide range of factors, some intrinsic to the munition, some to the agent it disseminates, and some, especially meteorological factors, to the environment. From the munition expenditure tables that are designed--on the basis of field trials--to enable field commanders to compute the number of munitions required for attacking particular targets, one may form an idea of the sorts of contamination density, in, say, kilograms of agent per hectare of target-area, with which a particular agent would in practice need to be spread in order to achieve those target effects by contact or by airborne challenge. Examples of such densities for different agents, standardized for meteorological conditions and expected target casualty-rates, are given in table 4.

Cost-effectiveness studies indicate that company-sized targets are optimal for nerve-gas attack. Against targets that are either smaller or larger, CW weapons would rarely ever be competitive with conventional or nuclear weapons. Under current tactical doctrines in Europe, a company-sized force deployed in the field would be expected to occupy maybe 20 or 25 hectares of ground. The expectation is that, in a central-European war, there would, on average, be around six such targets per kilometre of front at any given moment to a depth of 15-20 km. Those of the targets that were both acquirable by opposing weapons and of a kind that were significantly vulnerable to CW attack would be only a fraction of the total, maybe one-third.

Application of these rough rules of thumb to the figures given in table 4 suggests something of the size of the minimum supply of toxic agents that could have military significance in a European war. Nerve-gas stockpiles amounting to less than some hundreds of tons of agent could not, it would seem, be of any great significance.

An upper significance bound can also be envisaged, established by the maximum delivery capacity of all those deployed weapons that are capable of delivering CW agents. From what is publicly known of the number and firing-rates of CW-capable weapons deployed on either side of the central European confrontation, including those that might be assigned to deep-interdiction missions, and having regard to the fact that the weapons would be required for delivery of antimateriel as well as antipersonnel munitions, the upper bound would seem to be in the region of 700 - 1,000 agent-tons per day on either side: say 20 000 - 30 000 tons for a 30-day war.

As a rough guide, then, the scale of production of toxicants which should most concern us is on the order of 10^2 - 10^5 tons.

VI <u>Future CW weapons</u>[9]

The principal threat agents identified above have been in the arsenals for more than twenty years. In this sense, CW weaponry appears to be a mature technology, one in which such innovation as there has been since, say, the mid-1950s, has been marginal: incremental or evolutionary, rather than revolutionary. When set against the recent extraordinary rapidity of innovation in other areas of technology that exploit the molecular and life sciences, this seems surprising; and it may, indeed, betoken a great overhang of applicable scientific knowledge not yet reduced to CW practice. The relevant consideration, perhaps, is that major innovation requires major adjustment on the part of the organization that is to assimilate it. CW lies on the periphery of military theory and practice; under existing systems of military doctrine and organization, the incentive to draw CW any further into the mainstream is constrained by the weighty opportunity costs, chiefly organizational costs, of doing so.

The implication of this line of reasoning is that if, for one reason or another, the aforementioned incentive should grow, then the constraints inhibiting the CW exploitation of scientific knowledge acquired during the 1960s-1980s could fall away very rapidly, perhaps calling forth technological innovation in toxic-warfare methods that was revolutionary rather than incremental. One might point to the burgeoning biotechnologies as an area of industrial enterprise that might provide especially rich soil for the growth of new CW technologies.

If that implication is a valid one, then what lies before us may be a rapid upgrading of the status of toxic-weaponry in war-planning, bringing in its train competitive East-West armament in the form of a toxic arms race, and accelerating proliferation of toxic weapons in regions of the world so far spared the deployment of mass-destruction weapons. The incremental technological trends that characterize what we might then see as a pre-revolutionary present could perhaps come to furnish the absent incentive.

One of today's trends is currently manifesting itself in the so-called 'binary' munitions. These are loaded, not with actual toxicant, but with separate containers of relatively harmless reactants which are adapted to mix together to generate toxic agent during the period immediately preceding coincidence of the munition and its target: while the bomb is falling, for example, or after the artillery shell has been fired. Such munitions could present fewer hazards in production and storage, thereby lessening some of the opportunity costs of deploying toxic-warfare capabilities. In the West, this development is approaching the stage of large-scale production.

A binary munition is essentially an unactivated reactor for achieving the final stage of toxic-agent synthesis. Those binary munitions that are on the verge of entering the arsenals are designed to generate, not novel CW agents, but familiar old ones: sarin, VX and soman. The technology could, of course, lend itself to other agents also, for example, ones whose storage-stability had hitherto been judged too low for CW purposes.

A second trend is illustrated in the search that has been continuing since the earliest days of CW-weapons R&D for ways of breaching or otherwise circumventing antichemical protection. Of all the factors that are responsible for the present lowly status of CW in today's armamentaria, the most important is the relative ease in comparison with other weapons with which protection can be, and is, provided. That protection has, as its cornerstone, the physical processes of filtration. The presence of so much cyanogen chloride and hydrogen cyanide in the CW arsenals of World War II was largely due to the capacity of these small-molecule agents for saturating the charcoal filters of existing respirators (a vulnerability that has since been reduced by new charcoal impregnants). Quite a wide variety of other mask-penetrating and, more recently, clothing-penetrating concepts have been examined, but so far all of them have operational shortcomings. However, if a new concept did emerge that lacked those shortcomings, it might well have a deep effect on the overall security significance of CW weapons.

Weapon concepts can be envisaged which would seek to exploit other vulnerabilities in the overall array of equipments and procedures that afford antichemical production. But for the present one might judge that the significance of such weapons would be nowhere near as great as efficient mask- or clothing-penetrators. If, however, combat forces came to place more reliance than they do today on, for example, automated field alarms--if the physical protection of masks were erected only upon signals from automatic CW-agent detectors--then agents that did not trigger the alarms would grow in importance. Herein lies about the only plausible logic for lethal CW agents other than the nerve gases; and it is of course by no means inconceivable that agents competitive with the nerve gases could emerge. Here one thinks, in particular, of current talk about toxins and their potential availability from novel biotechnological processes. And there are the skin-penetrating carbamates to which the Swedish CD delegation has recently referred, the toxic organosilicon compounds, and various others. Yet, whether newly available from scientific research, or newly accessible in militarily significant quantities by virtue of developments in the chemical industry, the importance of such novel agents could, for the moment, hardly be more than marginal. Because the filtration processes that would be on hand to shield people from them are essentially physical in nature, they are independent of the precise chemical identity of the challenge.

A third trend lies in the further exploitation of casualty-incapacitant concepts. Within the category of incapacitating casualty agents, there is one family which, even now, could conceivably warrant as much concern as the mustard and nerve gases. These are the psychotropic glycollates. As casualty agents, they are inferior to the nerve gases, but the reasons for that inferiority can be seen to lie chiefly in particular shortcomings of the agents which could, in principle, be remedied. In terms of toxicity--in the sense of size of casualty-producing dose--the glycollates are comparable with the nerve gases in their potency, as the entry in table 4 for one of them, agent BZ, indicates. But they are solids at field temperatures, meaning a low propensity for establishing contact hazards; and, even by the inhalation route (when aerosolized), their onset of effects is delayed by tens of minutes. One remedy might be to use them in solution, the solvent chosen being one that was itself capable of exerting rapid toxic effects (such as a liquid irritant agent). Table 5 lists some of the more psychoactive of the glycollates, including ones that have been the more seriously considered for weaponization.

The glycollates are not alone in their candidacy as CW incapacitating casualty agents. Various other families of psychotropic drug have CW potential: benzomorphans, cannabinols, phenothiazines and others. So do a variety of toxins, including bacterial toxins, mycotoxins and other phytotoxins.

VII Proven large-scale routes to the principal CW agents[10]

Table 6 notes some of the manufucturing facilities that have been used in the past for large-scale production of today's principal CW agents. The purpose of the table is to indicate (a) the sizes of the factories posited by past military requirements for mustard and nerve gases, and (b) the production routes to those agents that have been favoured under past economic and technological circumstances. The table is not meant to be comprehensive. It makes no reference to, for example, French and Soviet production capacities for nerve gas, nor to the factories of other countries having experience of plant-scale mustard-gas production at the time of World War II (which include Italy, Hungary, Poland, the then Netherlands East Indies, Japan and South Africa).

Mustard gas

Probably by far the greatest proportion of the world's mustard gas has been produced by variants of the original process described by Despretz in 1822 and by Guthrie in 1860, namely the interaction of ethylene and chlorinated sulphur:

$$2CH_2{:}CH_2 + SCl_2 \longrightarrow S(CH_2CH_2Cl)_2$$

$$2CH_2{:}CH_2 + S_2Cl_2 \longrightarrow S(CH_2CH_2Cl)_2 + S$$

Much scope exists for varying the reaction conditions to alleviate problems presented by the numerous byproducts and side-products.

In one set of variants, the starting material is sulphur monochloride, as in the Levinstein Process that was used for all of the US production during World War II, for a minor proportion of the German production then ('direkt-Lost') and for much of the USSR's production. The US product (known as 'Levinstein mustard' or 'H') comprised no more than about 80 percent mustard; a part of the output was subjected to a process of water-washing and vacuum-distillation that yielded a product ('distilled mustard' or 'HD') of increased purity (ca. 96%) and much improved storage stability.

In a second set of Despretz-Guthrie variants, stemming from the methods used in France during World War I and subsequently, the starting material is sulphur dichloride. The technical grade of this is in fact an equilibrium mixture of dichloride, monochloride and free chlorine, which meant that in many of the processes studied the product displayed defects similar to the monochloride product due to sulphur precipitation and polysulphide formation. These problems were circumvented by the British during World War II by fractionation of the chlorinated sulphur over phosphorus pentachloride so as to provide, as starting material, a sulphur dichloride of high purity.

The Despretz-Guthrie method was one of two laboratory routes to mustard gas described during the nineteenth century. The second was the Meyer method, involving chlorination of thiodiglycol. Several adaptations of it have been developed and used for large-scale production. In terms of the number of countries that have used the Meyer method, it is probably the most widespread large-scale route to mustard gas, finding its most recent application in the mustard used by Iraq against Iran. The chlorinating agents that have been studied include phosphorus trichloride and thionyl chloride, but the most used appears to have been hydrochloric acid:

$$S(CH_2CH_2OH)_2 + 2HCl \text{ --------------> } S(CH_2CH_2Cl)_2 + 2H_2O$$

It was this that went into the 'Oxol' process of the German Reich, and into the second of the two British World War II processes for mustard. The objective in the British process was in fact an incomplete chlorination of the thiodiglycol so as to yield a product (known as 'runcol' or 'HT') that was rich in higher mustards, chiefly agent T:

$$2S(CH_2CH_2OH)_2 + 2HCl \text{ --------------> } O(CH_2CH_2SCH_2CH_2Cl)_2 + H_2O$$

The Meyer process has been adapted to plant-scale production of nitrogen mustards by using the appropriate _bis_- or _tris_(ethanol)amine in place of the thiodiglycol; thionyl chloride has been a preferred chlorinating agent:

$$2N(CH_2CH_2OH)_3 + 3SOCl_2 \text{ ------------> } 2N(CH_2CH_2Cl)_3 + 3H_2O + 3SO_2$$

There are other routes to mustard. Research in the USA during World War II demonstrated the feasibility of photosynthetic pathways. Hydrogen sulphide could be made to add to vinyl chloride under ultraviolet irradiation in the presence of a suitable catalyst and photosensitizer; and with appropriate mercaptans or chloroalkenes, a whole range of higher mustards was also accessible. It is not publicly known whether such methods have been operated on a large scale, but a CW textbook published in the GDR in 1967 has this to say: "Industrial photosynthesis....has been introduced in the countries which are the leading producers of CW agents as a part of the reorganisation of such plants".[11]

Tabun nerve-gas

At least three routes to tabun are described in the literature. The one used on a plant scale in Germany during World War II was a two-stage process starting from phosphoryl chloride and dimethylamine, the latter being produced on-site from methanol and ammonia:

$$2Me_2NH + POCl_3 \text{ -----> } Me_2N.P(O)Cl_2 + Me_2NH_2Cl$$

$$Me_2NH_2Cl + POCl_3 \text{ -----> } Me_2N.P(O)Cl_2 + 2HCl$$

$$Me_2N.P(O)Cl_2 + 2NaCN + EtOH \text{ -----> } Me_2N.P(O)(CN)OEt$$

It was conducted as a batch process, the plant being subdivided into twelve production units. Per 100 tons of tabun (produced in varying degrees of dilution with chlorobenzene), the raw-materials consumption was:

Ammonia, 100%	25 t
Methanol	60
Sodium cyanide	30
Phosphoryl chloride	134
Ethanol, absolute	37
Tabun	100 tons

Sarin/soman-family nerve gases

A wide variety of different routes are described in the literature for laboratory-bench synthesis of G agents in the sarin/soman family. The immediate precursors of the nerve gas in these routes include such species as:

$(R'O)_2PF$
$RP(O)X_2$ (where X = -Cl or -F)
$R'O.RP(O)Z$ (where Z = -Cl, $-NMe_2$, -SR or -SM)

Such precursors are themselves accessible by varied routes.

Only for sarin are the production routes that have actually been used for large-scale manufacture, or routes that have been developed to the point where they could have been so used, described in open sources. These routes may be grouped into two categories according to their key intermediate - diester or dichlor. Soman is presumably accessible via all the routes to sarin. But in view of the difficult accessibility of the pinacolyl alcohol needed to introduce its 1,2,2-trimethylpropyl group, diester processes would most likely not be favoured.

The diester processes that have been described for sarin went by way of diisopropyl methylphosphonate (DIMP), the final steps being:

$$({}^{i}PrO)_2MeP(O) + COCl_2 \text{ ------> } {}^{i}PrO.MeP(O)Cl$$

$${}^{i}PrO.MeP(O)Cl + HF \text{ ------> } {}^{i}PrO.MeP(O)F$$

The DIMP itself was the product of one or another of the then-available P-methylation reactions, the thermodynamics of which demand P(III) rather than P(V) precursors. In the USA, two reactions were developed for the purpose during the early 1950s, the 'US Salt Process' and the 'TIPP Process', these corresponding to the two classical methods for forming carbon-phosphorus bonds: the Michaelis-Becker reaction, from the sodium salt of diisopropyl hydrogen phosphite, and the Arbuzov reaction, from triisopropyl phosphite.

In the other category of plant-scale routes to sarin, the production methods went by way of methylphosphonyl dichloride (dichlor), as in Gerhard Schrader's original process:

$$MeP(O)Cl_2 + NaF + {}^{i}PrOH \text{ ----------> } {}^{i}PrO.MeP(O)F$$

As subsequently developed in Germany, the intermediate methylphosphonyl difluoride (otherwise known as 'difluor' or 'DF') was isolated in the form of a dichlor/difluor mixture ('didi') and then esterified:

$$2MeP(O)Cl_2 + HF \text{ -------> } MeP(O)Cl_2 + MeP(O)F_2 + 2HCl$$

$$\text{Didi} + {}^{i}PrOH \text{ -------> } {}^{i}PrO.MeP(O)F$$

It was this didi process that was later used, during 1952-57, to produce the present US stockpile of sarin. And it is this process, in modified form, that is used in the leading US binary munition. The modification lies in esterifying, not a didi mixture, but difluor alone, produced by fluorination of dichlor. To get the esterification to proceed with sufficient rapidity in the binary munition, the isopropanol loading contains some 28 percent of isopropylamine to mop up byproduct hydrogen fluoride.

Of the dichlor consumed in the US sarin programme, the first 200 tons came from plant exploiting, by what was known as the 'APC Process', the Perren-Kinnear reaction for carbon-phosphorus bond formation: water was used to break the complex formed by interaction of methyl choride, phosphorus trichloride and anhydrous aluminium chloride. This was, however, a stop-gap measure until the huge dichlor plant built for the purpose on the Tennessee Valley Authority reservation finally came on stream. The plant exploited the second of the dichlor production methods that had been developed in wartime Germany, namely chlorination of the product of thermal rearrangement of dimethyl hydrogen phosphite (DMHP). (Schrader's original method, subsequently used on plant-scale at Dyhernfurth, was a Michaelis-Becker--'salt'--process going by way of the sodium salt of DMHP.) The thermal-rearrangement product, known as 'pyromix', was essentially a mixture of dimethyl methylphosphonate and the corresponding pyrophosphonate. The chlorinating agent used was phosphorus trichloride and chlorine, giving rise to large quantities of byproduct phosphoryl chloride.

The raw-materials requirements estimated for the German DMHP/dichlor/didi process were as follows:

phosphorus trichoride	510 tons
chlorine	145
methanol	160
isopropanol	58
hydrogen fluoride	20
sarin	100 tons

The US process presumably had a rather similar consumption. Something like two-thirds of the input phosphorus trichloride would have served as chlorinating agent rather than as the P(III) source. By the time US production of sarin was drawing to a close, a route to dichlor much more economical of phosphorus trichloride (and therefore of the elemental phosphorus from which it was made) had been developed. This was the 'HTM Process', in which methyldichlorophosphine, produced by high-temperature methanation of phosphorus trichloride in the presence of catalytic quantities of oxygen, was used together with chlorine to chlorinate the pyromix, in the process itself becoming oxidized to dichlor.

Methyldichlorophosphine can, of course, readily be oxidized to dichlor with oxidants other than chlorine/DMHP-pyromix, for example by oxygen itself. Had the technology become available earlier, US dichlor production would perhaps have rested on the HTM Process alone. Competing with it was the 'ASP Process' for methyldichlorophosphine production, under development in the British CW programme: an immensely hazardous route involving methyl and dimethyl aluminium chlorides.

All of the publicly described routes to sarin-family G agents have started from phosphorus trichloride, all of which was made by chlorination of elemental (white) phosphorus. P(V) starting materials, such as phosphorus pentoxide or phosphoryl chloride, were ruled out by the requirements of the methyl-phosphorus bond formation. Access to elemental phosphorus was thus crucial for sarin production.

Large-scale phosphorus plants, by virtue of the immense power consumptions involved, are rather few in number in the world. Of relevance here, therefore, are the various emergent technologies for producing methyldichlorophosphine or dichlor directly from phosphate rock without having to isolate elemental phosphorus. Also relevant, for the same reason, are the emergent technologies for producing elemental phosphorus by routes other than the conventional electrothermal processes.

VX and other V agent nerve gases

At least eight different routes are described in the literature for laboratory-bench syntheses either of V agents or of their close congeners. The immediate precursors of the nerve gas (or of its readily convertible thiono isomer) in these routes include such species as:

$R'O.RP.OCH_2CH_2NR''_2$

$R'O.RPX$ (where X = -OH or -OR')

$R'O.RP(O)Z$ (where Z = -Cl or -SM)

$R'O.RP(S)Cl$

$R'O.RP(O)SCH_2CH_2Y$

$Cl.RP(O)SCH_2CH_2NR''_2$

It was the first of these which came to be used in the only plant-scale process thus far described in the open literature, namely the one run by the US Army during 1961-67 to produce the present US stockpile of VX.

That process--the 'Newport Process'--had, as its final stage, sulphidation of the mixed P(III) ester known as 'QL':

$EtO.MeP.OCH_2CH_2NPr^i_2 + S$ ------> $EtO.MeP(S)OCH_2CH_2NPr^i_2$

$EtO.MeP(S)OCH_2CH_2NPr^i_2$ ------> $EtO.MeP(O)SCH_2CH_2NPr^i_2$

The heat of the sulphidation reaction was sufficient to drive the thiono-thiolo isomerization. The QL had been produced (and isolated) during the preceding step by transesterification of diethyl methylphosphonite, which itself had been produced from methyldichlorophosphine made in an on-site HTM plant.

$$PCl_3 + CH_4 \text{ -----> } MePCl_2$$

$$MePCl_2 + 2EtOH \text{ -----> } MeP(OEt)_2$$

$$MeP(OEt)_2 + HOCH_2CH_2NPr^i_2 \text{ -----> } QL + EtOH$$

Of the various alternatives to this transesterification route that were studied during the development work, what appear to have been the principal competitors all used the aminoethyl chloride to introduce the basic-ester group. Either the chloride was reacted with an intermediate that had already undergone the requisite sulphidation:

$$M[MeP(O)(S)OEt] + R''_2NCH_2CH_2Cl \text{ ------> V agent} + MCl$$

$(M = NH_4$, K or Na)

or it was reacted with the appropriate P(III) intermediate simultaneously with sulphur:

$$EtO.MePOH + ClCH_2CH_2NR^i_2 + S \text{ ------> V agent}$$

In all the routes, the phosphorus moiety originated in alkyldichlorophosphine: $MePCl_2$ for VX and its P-methyl congeners, or $EtPCl_2$ for the P-ethyl homologues, such as agent VE.

The two main systems that have been developed for binary VX munitions have both closely followed the Newport Process. One uses an identical reaction, QL being one of the two binary reactants, the other being finely divided rhombic sulphur. The second binary system uses a liquid dimethyl polysulphide in place of the sulphur.

Quinuclidinyl benzilate

So far as is known, only one glycollate incapacitant has ever been put into anything approaching large-scale production for CW purposes, namely 3-quinuclidinyl benzilate, otherwise known as agent BZ. About 50 tons were manufactured during 1963-64 by a US corporation under US Army contract. The corporation's own plant was apparently used.

The process used appears to have been based on a variant of one of the two laboratory methods described in the literature, namely transesterification of the sodium derivative of methyl benzilate with 3-quinuclidinol:

$Ph_2C(ONa)CO_2Me$ + QOH -----> $Ph_2C(OH)CO.OQ$

where Q = 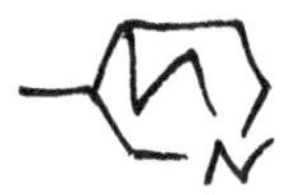

VIII The major CW-agent precursors

Listed in table 7 are thirty of the intermediates referred to in section VII that have been used, or been developed for use, in plant-scale production processes for mustard gas, tabun, sarin, VX and BZ. The table identifies which of these four CW agents the chemicals listed are intermediates for, and how many production steps they are away from the agents.

As has been seen, all four agents are members of larger chemical families, meaning that congeners differing only slightly in chemical structure might also serve as CW agents. So, homologues of some of the intermediates listed in table 7 may, in certain cases, also need to be regarded as significant CW-agent precursors. The following comments illustrate the point:

a) Powerful incapacitants in the glycollate family can be made by substituting for either of the BZ-precursors listed in table 7. Thus, reference to table 5 suggests that, if 3-quinuclidinol is to be included in a list of major CW-agent precursors, then so too should such other alcohols as N-methyl-4-piperidinol; and if methyl benzilate is to be included, then so too should other glycollate esters. The case for doing so is all the stronger since the agents listed alongside BZ in table 5 include ones that are among the candidates under evaluation as replacements for BZ, which is now considered obsolete.

b) If a decision were being made today on which of the many candidate V agents to select for weaponization, VX might very well not be the agent of choice. Its O-cyclopentyl homologue, for example, might be preferred (on grounds of refractoriness to oxime therapy)--in which case the O-cyclopentyl homologues of intermediates nos 11 and 17 should also be listed. No doubt the same could be said of other O-alkyl homologues. Nor is VX so greatly superior as a CW agent (especially for binary-munition purposes) to V agents having different N-substituents that exclusion of intermediates homologous to nos 17, 27 or 28 would be justifiable. The same is true of the P-ethyl series of V agents, meaning that the P-ethyl homologues of nos 11, 16, 17, 18, 19 and 20 should not be excluded. Nor--to emphasize this matter of substitution still further--might it be entirely justifiable to exclude the selenium analogues of the V agents from consideration.

c) Similar remarks hold for the sarin family of G agents: the O-alkyl and P-ethyl homologues particularly. For some of the former, notably soman, the alcohols that would have to be used in any production could be sufficiently peculiar to warrant listing as significant precursors--pinacolyl alcohol, for instance, or even such substances as 2-chlorocyclopentanol. Again, reference to table 3 suggests that precursors corresponding to the thionofluoridate or thionoazidate analogues of sarin-family agents might merit inclusion; but very little information exists about these analogues in the open literature.

d) The tabun family of G agents seem to present fewer serious options for substitution. The O-isopropyl homologue appears to exceed tabun in toxicity, but no 8 would still be its precursor. There seems to be nothing in the open literature to suggest that any of the NN-homologues would be superior to tabun.

Table 7 is a reflection of chemical-process technology as it existed 25-45 years ago. The production methods that might be favoured nowadays, or within the newly developing chemical economies, could well be different. Some immutabilities can, nonetheless, be identified. For example, the great strength of the carbon-phosphorus bond in P-alkyl compounds will surely continue to mean that unimpeded access to supplies of methylphosphonites or methylphosphonates (or the P-ethyl homologues) would significantly ease the task of acquiring a militarily significant nerve-gas capability.

Let us, then, in the light of the foregoing considerations, transform the listing of precursors contained in table 7 into a version better focussed on today's situation. Since, in several cases, the specification of individual precursors is clearly invidious, let us instead specify, where appropriate, families of homologues. And let us then apply a weighting so as to develop a rank-order of CW significance, using that weighting not only for the precursors but also for industrial chemicals that are potential CW agents.

Such a weighting ought to reflect at least two factors. One is the closeness of a particular precursor or precursor-family to its derivable CW agent. The numbers in the right-hand columns of table 7 can serve for this purpose. For chemicals that are themselves potential CW agents, the assignable weighting factor would be 1. The second factor is the relative military significance of the CW agents concerned. This is more difficult. Although attempts have been made (as noted earlier) to rank CW agents according to their CW utility on objective criteria,[6] these have been only partly successful: there are just too many pertinent considerations for the judgement not to be highly subjective. In table 8 a rank-order is proposed. The rank numbers shown there can serve as the second weighting factors.

The conclusion towards which this paper has been working is now within reach, for, with these weighting factors, a basis has been developed for ranking industrial chemicals according to the degree of attention that ought to be paid to them in the design of an international CW-weapon nonproduction regime: chemicals that may themselves serve as CW agents or which may be used to produce CW agents either in industrial plant or in binary munitions. The conclusion is shown in table 9.

IX Correlating categories of chemical with categories of control

To rank chemicals according to their significance to the purpose of the CWC is not at all the same thing as ranking chemicals according to the degree of stringency with which they should be controlled in order to make the CWC regime an effective one. The point is well illustrated by the presence of ethylene and vinyl chloride in Rank 3 of table 9 (at the same level as an immediate precursor of the nerve-gas tabun). These chemicals are in the bedrock of the chemical industry, produced in enormous volume all around the world for a great variety of beneficial applications. Notwithstanding their possible application in mustard-gas production, any attempt to control them would simply bring the CWC regime into ridicule.

This circumstance can be seen as a specific embodiment of a general proposition. Participation in a CWC regime will be at some cost to each state party: there will be financial, administrative and other such burdens to be borne, even political costs to the country's leadership. The higher such costs, the more disinclined may governments be to see greater benefit in joining the regime than in remaining outside it; and, even if they did join, subsequent pressure to cut costs could subvert the proper working of the regime. Either way, the regime would lose its effectiveness as the instrument for implementing the purpose of the CWC. To attempt to control ethylene would be to incur heavy administrative costs. And the antagonism of the chemical industry that could result (whether in a market or a centrally planned economy) could constitute a serious political cost.

In this regard, one may see ethylene and vinyl chloride as being at one end of a spectrum of industrial sensitivities. Towards the other end are such chemicals as the alkyl phosphites for which there may be only one producer in a given country. An obligation upon a government to declare production figures and other data might then require it to broadcast proprietary information which the enterprise concerned (or its customers) could find highly disadvantageous. In between, there are other sensitivities whose nature it is one of the purposes of this conference to clarify.

These considerations indicate that a further process of analysis must be interposed between a ranking developed, as above, from security principles and the transformation of that ranking into categories of chemical defined according to the stringency of the controls that are to be applied to them. That intermediary process must somehow introduce the realities of the chemical-industrial world into the formulation of control requirements.

A conceptual tool for doing this is to treat the security ranking as one dimension of a matrix. Along the second dimension would be a categorization of chemicals done in terms of their industrial status: a classification such as is outlined in table 10. Individual chemicals or families of chemicals would then be entered into each cell of the matrix. Once that was done, one would acquire the necessary synoptic view of the options for aggregating or disaggregating the security categories according to the controls which it might be feasible to apply to each one. That would be an elaborate exercise, and is not attempted here. Its main virtue would lie in the basis it could offer for structuring technical negotiations of some delicacy.

Correlations of categories of control-regime with categories of chemical exemplified with lists of particular chemicals have been put forward in three plenary-CD working papers: the US draft CWC of 18 April 1984 (CD/500), the British paper of 10 July 1984 (CD/514) and the Swedish paper of 20 August 1985 (CD/632). It is instructive, in the light of the foregoing, to compare their rankings of chemical. This is done in table 11.

How the process of reconciling alternative approaches to the correlation will develop in the future is a matter for conjecture--most uncertain conjecture at that, given the conspicuous silence of some important voices. Maybe it is premature to suppose that the need to accommodate industrial interests, especially those in the market economies, will prove rate-determining. Some Western governments acted with remarkable rapidity during mid-1984 in gaining industry acceptance of special export controls on several of the chemicals of concern to the CWC.

These actions were stimulated by the UN Secretary-General's verification of the use of chemical weapons in the Gulf War, and by the accumulating evidence that Iraq had been acquiring its CW capability in part through the assistance, presumably unwitting, of certain commercial enterprises in the Western world.

A summary of those Gulf-War-related export controls is given in table 12. For some countries, they were simply augmentations of controls already in place expressly for the purpose of inhibiting chemical-weapons technology-transfer. Those of the USA are noted in the Appendix, where lists of the controlled chemicals are reproduced and analysed.

The degree of government-industry cooperation in the field of CW constraints which the various export-control regulations represent seems impressive; and intergovernmental negotiations, largely unpublicized, were set in motion during 1985 to broaden the controls.[12] This is a foundation that can surely be built upon in the further elaboration of the projected Chemical Weapons Convention. What ultimately is needed is not a system that will guarantee the nondiversion of each and every pertinent chemical from industrial applications into CW capability. It is instead a system which, by focussing on an appropriate range of key chemicals, will sustain confidence in the overall CWC regime.

Table 1. Toxicants that have been used or stockpiled as antipersonnel CW agents[4b]

Initial	Toxicant and its CW history (A: Stockpiled today. B: Post-WW2 use. C: Stockpiled in WW2. D: Heavy WW1 use. E: Minor WW1 use)	
HARASSING INCAPACITATING AGENTS: TEAR GASES AND OTHER SENSORY IRRITANTS		
1914	E	. . . ethyl bromoacetate
1914	E	. . . o-dianisidine chlorosulphonate
1914	E	. . . chloroacetone
1915	D	. . xylyl bromide
1915	E	. . . xylylene bromide
1915	D	. . benzyl bromide
1915	E	. . . capsaicin
1915	E	. . . methyl chlorosulphonate
1915	E	. . . ethyl chlorosulphonate
1915	E	. . . chloromethyl chloroformate
1915	E	. . . dichloromethyl chloroformate
1915	D	. . bromoacetone
1915	E	. . . bromomethylethyl ketone
1915	E	. . . iodoacetone
1915	E	. . . dimethyl sulphate
1915	CD	. ethyl iodoacetate
1915	E	. . . benzyl iodide
1915	E	. . . o-nitrobenzyl chloride
1915	E	. . . benzyl chloride
1916	E	. . . acrolein
1917	CD	. diphenylchloroarsine [DA]
1918	E	. . . thiophosgene
1918	DC	. diphenylcyanoarsine [DC]
1918	E	. . . N-ethylcarbazole
1918	CD	. 1-bromo-1-tolunitrile [CA]
1918	E	. . . phenyldibromoarsine
1918	ABCE	10-chloro-5,10-dihydrophenarsazine [adamsite] [DM]
1919	ABC	1-chloroacetophenone [CN]
1958	AB	2-chlorobenzalmalononitrile [CS]
1974	A	dibenzoxazepine [CR]
LETHAL AGENTS: CHOKING GASES		
1915	D	. . chlorine
1915	E	. . . bromine
1915	E	. . . perchloromethylmercaptan
1915	A(?)CD	phosgene
1916	CD	. trichloromethyl chloroformate [diphosgene]
1916	CD	. chloropicrin
1917	E	. . . phenylcarbylamine chloride
1918	E	. . . bis(chloromethyl) ether
1918	E	. . . bis(bromomethyl) ether
LETHAL AGENTS: BLOOD GASES		
1916	A(?)CD	hydrogen cyanide
1916	E	. . . hydrogen sulphide
1916	E	. . . cyanogen bromide
1916	CD	. cyanogen chloride
1918	E	. . . cyanoformate esters
LETHAL AGENTS: BLISTER GASES		
1917	ABCD	bis(2-chloroethyl) sulphide [mustard gas]
1917	CD	. phenyldichloroarsine
1918	E	. . . ethyldichloroarsine
1918	E	. . . methyldichloroarsine
1918	E	. . . ethyldibromoarsine
1918	A(?)B(?)C	2-chlorovinyldichloroarsine [lewisite]
1938	AC	bis(2-chloroethylthioethyl) ether
1940	A(?)C	tris(2-chloroethyl)amine
LETHAL AGENTS: NERVE GASES		
1942	AB(?)C	ethyl NN-dimethylphosphoramidocyanidate [tabun]
1952	A	O-isopropyl methylphosphonofluoridate [sarin]
1961	A	O-ethyl S-2-diisopropylaminoethyl methylphosphonothiolate [VX]
c 1962?	A(?)	O-1,2,2-trimethylpropyl methylphosphonofluoridate [soman]
OTHER LETHAL AGENTS		
1943	C	botulinal toxin
CASUALTY INCAPACITATING AGENTS		
1963	A	3-quinuclidinyl benzilate [BZ]

Table 2. Other CW threat agents

	Toxicants and other chemicals whose quantity production for CW purposes has, at one time or another, been seriously considered, suspected or alleged
I.	Arsine (or metal arsenides)
	Dichloroformoxime (phosgene oxime)
II.	Ricin
	Shellfish poison (saxitoxin)
	Staphylococcal enterotoxin B
	T2 mycotoxin
	DAS mycotoxin (anguidine)
	Aconitine
III.	Phencyclidine
	[Staphylococcal enterotoxin B]
IV.	Chlorine trifluoride
	Iron pentacarbonyl
	Perfluoroisobutene
	[Dichloroformoxime]
V.	Certain bisquaternary ammonium compounds
	[Ricin]
	[Saxitoxin]
	Certain carbamates
	[Aconitine]
VI.	Certain silatranes
	Certain trioxaphosphabicyclooctanes
	Tetrachlorodibenzo-p-dioxin

I: Of special concern during World War II.

II: Toxins.

III: Incapacitants approved for weaponization.

IV: Considered for their filter-breaking properties.

V: Considered for direct-injection toxic ammunition.

Table 3. The main nerve-gas families[7]

I	$R'O(R_2N)P(O)CN$	R, R' and R" denote alkyl groups that may be the same or different.
II	$R'O.RP(O)F$	
III	$R'O.RP(S)F$	
IV	$R'O.RP(S)N_3$	
V	$R'O.RP(O)OR'$	
VI	$R'O.RP(O)SCH_2CH_2NR''_2$	

Note: There are several other families of organophosphorus anticholinesterase agent for which a degree of CW potential exists. They include ones which have in the past found agrochemical or medicinal applications, such as:

Dyflos	$(^iPrO)_2P(O)F$
Paraoxon	$(EtO)_2P(O)OC_6H_4.NO_2^p$
Amiton	$(EtO)_2P(O)SCH_2CH_2NEt_2$
Armin	$EtO.EtP(O)OC_6H_4.NO_2^p$

A US defence laboratory reported in 1973 that "a class of compounds has been uncovered that shows promise as intermediate volatility [nerve] agents". Whether that class, whose chemical identity was not disclosed, falls within the six families here identified is not known.

Table 4. Contamination densities prescribed for different CW agents[13]

Agent	Contact hazard	Airborne hazard
	(tons per square kilometre)	
Phosgene	---	21
Hydrogen cyanide	---	26
Mustard gas	19	4
Tabun	14	2
Sarin	..	0.5
VX	2	..
BZ incapacitant	---	0.6
CS tear gas	---	0.7
24-D (to kill rice plants)	1.2	---

Note: For the casualty agents listed (i.e., all except CS and 24-D), the figures are for 50 percent casualty rates among unprotected target populations, and for a standard set of meteorological conditions (typical of a warm, overcast day).

Table 5. Some highly psychoactive glycollates

Identifier	Chemical structure RO.CO.C(OH)(R')R"			Literature reference
	R	R'	R"	
BB	N-methyl-3-piperidyl	phenyl	phenyl	a
..	N-methyl-3-piperidyl	phenyl	cyclopropyl	a
..	N-methyl-3-piperidyl	phenyl	cyclobutyl	a
..	N-methyl-3-piperidyl	phenyl	cyclopentyl	a
..	N-methyl-3-piperidyl	phenyl	cyclohexyl	a
PMCG	N-ethyl-2-pyrrolidylmethyl	phenyl	cyclopentyl	a
..	N-methyl-3-pyrrolidyl	phenyl	cyclopentyl	a
..	N-methyl-4-piperidyl	phenyl	cyclopentyl	c
PCMG; 302196	N-methyl-4-piperidyl	1-propynyl	cyclopentyl	f
EA 3834	N-methyl-4-piperidyl	phenyl	isopropyl	
EA 3580	N-methyl-4-piperidyl	phenyl	cyclobutyl	
..	N-methyl-4-piperidyl	phenyl	phenyl	b, d
BZ	3-quinuclidinyl	phenyl	phenyl	b, d
..	3-quinuclidinyl	phenyl	2-thienyl	d, e
..	3-quinuclidinyl	2-thienyl	2-thienyl	d, e
..	3-quinuclidinyl	3-thienyl	3-thienyl	d

Literature references:

a L G Abood & J H Biel, Int Rev Neurobiol 4: 217-73 (1962)
b N W Gabel & L G Abood, J Med Chem 8: 616-19 (1965)
c L G Abood in A Burger (ed), Drugs Affecting the Central Nervous System, vol 2 (1968), pp 127-67.
d L Albanus, Acta Pharmacol et Toxicol 28: 305-26 (1970)
e K A Zaitseva et al, Farmakol i Toksikol 37(6): 671-74 (1974)
f A M Van der Poel, Psychopharmacologia 37: 45-58 (1974)

Table 6. Some representative former production facilities for major CW threat agents[13]

Country	Location	Production route	Production capacity (t/m)	Remarks
MUSTARD GAS				
Britain	Runcorn	Dichloride	1944: 1920	Further dichloride-route capacity existed at Salwick and Rhydymwyn, so that the total capacity then was about 4100 tons per month of this 'pyro' mustard.
Britain	Rhydymwyn	Thiodiglycol	1944: 440	Further thiodiglycol-route capacity for 'runcol' mustard existed at Runcorn, making the overall British mustard capacity then about 5,400 tons per month.
Canada	Cornwall	Thiodiglycol	1942: 1200	Actual production totalled about 5,000 tons.
France	Sorgues	Dichloride	1939: 240	Additional mustard capacity existed then at Angouleme.
Germany	Gendorf	Monochloride	1944: 4000	Actual production here, Germany's only 'direkt-Lost' plant, totalled about 3,000 tons.
Germany	Ammendorf	Thiodiglycol	1944: 700	Some 22,000 tons of mustard were made here, which was one of three 'Oxol-Lost' plants in Germany, those at Hüls and Gendorf providing a further 1,400 t/m capacity from which, however, the total output was only about 250 tons.
Soviet Union	Novomoskovsk	Monochloride	1941: 750	This, the oldest of the Soviet mustard plants, provided the model on which, according to German sources, some 17 others were built during the decade preceding the Great Patriotic War, most of them apparently using a Levinstein-type process. The USSR has, however, published no information on the subject. In 1945, Germany estimated the total Soviet mustard capacity to be 10,000 to 12,500 t/m, about double what it had been five years earlier when actual production was thought to have run at 10-12 percent of capacity.
United States	Pine Bluff	Monochloride	?	The three other large-scale US mustard factories, all of them also using the Levinstein process, were at the Edgewood, Huntsville and Rocky Mountain (Denver) arsenals. Total US mustard production during World War II was 92,000 tons.
TABUN				
Germany	Dyhernfurth	Phosphoryl chloride	1944: 800	Total German production of tabun, all of it at this plant, was about 12,000 tons.
SARIN				
Germany	Falkenhagen	Didi/thermal-DMHP	[1945: 500]	Unfinished. A smaller (100 t/m) sarin plant was built at Dyhernfurth using the original Schrader route, but it, too, did not go into production.
United States	Denver	Didi/Thermal-DMHP	1957: 2500	Total US Sarin production was about 13,000 tons, all of it here at Rocky Mountain Arsenal. The dichlor intermediate was made at Muscle Shoals, Alabama.
AGENT VX				
United States	Newport	Transesterification	?	Total US production of VX, all at this plant, was about 5,000 tons.

Table 7. Proven CW-agent precursors

Precursor	Production steps away from: H	GA	GB	VX	BZ
1) sulphur monochloride	1	-	-	-	-
2) sulphur dichloride	1	-	-	-	-
3) thiodiglycol	1	-	-	-	-
4) 2-chloroethanol	2	-	-	-	-
5) ethylene oxide	2	-	-	-	-
6) ethylene	1-2	-	-	-	-
7) vinyl chloride	1	-	-	-	-
8) dimethylphosphoramidic dichloride	-	1	-	-	-
9) methylphosphonyl difluoride	-	-	1	-	-
10) isopropyl methylphosphonochloridate	-	-	1	-	-
11) ethyl methylphosphonothioate salts	-	-	-	1	-
12) methylphosphonyl dichloride	-	-	1-2	-	-
13) dimethyl methylphosphonate	-	-	2-3	-	-
14) dimethyl pyromethylphosphonate	-	-	2-3	-	-
15) diisopropyl methylphosphonate	-	-	2	-	-
16) diethyl methylphosphonothionate	-	-	-	2	-
17) diisopropylaminoethyl ethyl methylphosphonite	-	-	-	1	-
18) diethyl methylphosphonite	-	-	-	2	-
19) ethyl hydrogen methylphosphonite	-	-	-	1	-
20) methyldichlorophosphine	-	-	2-3	2-3	-
21) dimethyl hydrogen phosphite	-	-	3-4	-	-
22) diisopropyl hydrogen phosphite	-	-	3	-	-
23) triisopropyl phosphite	-	-	3	-	-
24) phosphorus trichloride	-	3	2-5	3-4	-
25) phosphoryl chloride	-	2	-	-	-
26) dimethylamine	-	2	-	-	-
27) 2-diisopropylaminoethanol	-	-	-	2	-
28) 2-diisopropylaminoethyl chloride	-	-	-	1	-
29) methyl benzilate	-	-	-	-	1-2
30) 3-quinuclidinol	-	-	-	-	1-2

Key: H = mustard gas, GA = tabun, GB = sarin, BZ = 3-quinuclidinyl benzilate.

Table 8. A suggested ranking of CW agents according to their military significance

Rank	CW agent
1	Sarin/soman family of G agents
2	V agents
3	Tabun family of G agents Mustard gas family
4	BZ family of incapacitants Hydrogen cyanide Cyanogen chloride
5	Phosgene
6	Lewisite and other dichloroarsines Diphosgene Chloropicrin
7	Sensory irritants (adamsite, CS, CR, CN, etc)

Table 9. Chemicals ranked according to their significance for the regime to be established by the projected Chemical Weapons Convention

Rank	Chemical
1	Methyl- and ethyl-phosphonyl difluorides
	Alkyl methyl- and ethyl-phosphonochloridates
	Pinacolyl alcohol
2	Methyl- and ethyl-phosphonyl dichlorides
	Methyl- and ethyl-dichlorophosphines
	All other derivatives of methyl- and ethyl-phosphonic acids
	All other derivatives of methyl- and ethyl-phosphinic acids
	Mono- and di-alkyl esters of methyl- and ethyl-phosphonothioic acids
	The NN-dialkylaminoethyl alcohols, thiols and halides
3	Di- and tri-alkyl esters of phosphorous acid
	Phosphorus trichloride
	Dimethylphosphoramidic dichloride
	Thiodiglycol
	Sulphur mono- and di-chlorides
	Ethylene
	Vinyl chloride
	Hydrogen cyanide
	Cyanogen chloride
4	Alkyl arylglycollates
	N-substituted 3- and 4-piperidinols
	Phosgene
5	2-Chloroethanol
	Phosphoryl chloride
	Dimethylamine
	Ethylene oxide

Note: Single-purpose CW agents have not been included.

Table 10 A categorization by industrial status of chemicals significant to the Chemical Weapons Convention

Category definition	Examples of chemicals within the category having the following CW applications:			
	CW agents	Adjuvants of CW agents	Precursors of CW agents	
			Binary loading	Plant-process only
Chemicals without civilian applications other research	Soman		$MeP(O)F_2$	
	Lewisite		Pinacolyl alcohol	
Chemicals whose former civilian applications have now essentially vanished	Amiton			
Chemicals having civilian applications satisfied by annual worldwide production*				
of less than 1 000 tonnes	Some drugs	Some furans**		
in the range 1 000 to 10 000 tonnes	Some pesticides	DMSO** Carbodiimides**		$MeP(O)(OMe)_2$**
in the range of 10 000 to 100 000 tonnes	Tetraethyl lead**	Certain polymers**		$POCl_3$
of greater than 100 000 tonnes	HCN		$^{i}PrOH$	PCl_3
Chemicals of high toxicity produced on a substantial scale for military applications only, but ones unrelated to CW	Examples: some rocket propellants			
Chemicals having a large market potential which, if realised, would induce hitherto absent large-scale civil production capacity for intermediates having CW relevance	Examples: methylphosphonate pesticides			

* The four ranges specified below have been chosen arbitrarily and need to be considered further. The numbers eventually used should somehow reflect significant differences in numbers of producers. An altogether different approach that might be used would focus, not on production volumes or numbers of producers, but on categories of end-user. For the purpose of the CWC regime, end-use monitoring of divertible chemicals may be at least as important as production monitoring. Another possibly useful differentiator, then, would be the ratio of numbers of producers to numbers of end-use applications.

** Use of these chemicals to exemplify the category is <u>not</u> based on firm knowledge of its annual production.

Table 11 American, British and Swedish proposed correlations of chemicals and control regimes

Chemical to be controlled	CD/500 USA	CD/514 UK	CD/632 Sweden
Supertoxic lethal chemicals, inc sulphur mustard		Cat H	
-- tabun	Sch A		Reg I
-- sarin	Sch A		Reg I
-- soman	Sch A		Reg I
-- agent VX	Sch A		Reg I
-- amiton			Reg I
-- skin-penetrating carbamates			Reg I
-- saxitoxin	Sch A		
-- strophantin			Reg I
-- mustard gas	Sch A		
nitrogen mustards		Cat H	
lewisite		Cat H	
glycollate incapacitants		Cat H	
-- agent BZ	Sch A		Reg I
adamsite			Reg I
pinacolyl alcohol	Sch A&C	Cat H	Reg I
methyl-phosphorus compounds	Sch C	Cat H	
-- methylphosphonyl difluoride	Sch A		Reg I
-- NN-diisopropylaminoethyl ethyl methylphosphonite (QL)			Reg I
ethyl-phosphorus compounds	Sch C	Cat H	
propyl-phosphorus compounds	Sch C	Cat H	
NN-disubstituted-aminoethyl alcohols, thiols and halides	Sch C	Cat M	
-- NN-diisopropylaminoethyl alcohol, thiol and halides		Cat H	
methyl & ethyl esters of phosphorous acid	Sch C		
-- dimethyl hydrogen phosphite		Cat H	
-- diethyl hydrogen phosphite		Cat H	
-- triethyl hydrogen phosphite		Cat H	
phosphorus trichloride	Sch B	Cat M	Reg III
thiodiglycol	Sch B	Cat H	Reg II
sulphur monochloride		Cat M	
hydrogen cyanide	Sch B	Cat M	Reg III
phosgene	Sch B	Cat M	Reg III
arylglycollic acids		Cat H	
-- phenylglycollic acids	Sch C		
arylglycollates		Cat H	
cycloalkylglycollic acids	Sch C	Cat H	
cycloalkylglycollates		Cat H	
alkylglycollic acids	Sch C	Cat H	
alkylglycollates		Cat H	
3- and 4-piperidinol	Sch C	Cat M	
3- and 4-piperidinol derivatives	Sch C		
quinuclidinols		Cat M	
phosphoryl chloride	Sch B	Cat M	Reg III
arsenous chloride		Cat H	
chloropicrin	Sch B		

Table 12 Countries imposing new export controls on particular chemicals following the verification of the Iraqi use of CW

Significance of the chemical(s) as precursor for CW agent(s)

Countries[a] where governmental controls have been placed on export of the chemical(s)
□ bound for any destination, or
▨ bound for Iraq or Iran only

EC countries imposing further controls: NL, DK, GB, IRL

Controlled chemical(s)	Derivable CW agent(s)	No. of production steps away from it (them)	Availability of alternative precursors[c]	AUS	CAN	USA	JAP	EC[b]	NL	DK	GB	IRL
2-Chloroethanol	Mustard gas	2	Yes	□				□	□		▨	
Dimethylamine	Tabun nerve gas	2	No	□				□	□		▨	
Potassium fluoride	Sarin family nerve gases	1	Yes	□	□[d]	▨	▨	□	□		▨	
Thiodiglycol	Mustard gas	1	Yes	□	□	▨	▨	□	□	□	□	□
Phosphoryl chloride	Tabun family	2	No	□	□	▨	▨	□	□	□	□	□
Methylphosphonyl difluoride	Sarin family nerve gases	1	Yes	□	□	▨	▨	□	□	□	□	□
Dimethyl methylphosphonate	Sarin family nerve gases	2–3	Yes	□	□	▨	▨	□	□	□	□	□
Methylphosphonyl dichloride	Sarin family nerve gases	2	Yes	□			▨	□	□	□	□	□
All P-methyl and P-ethyl compounds	All nerve gases except the Tabun family	0–4	No						□			□
Methyl and ethyl phosphites	All nerve gases except the Tabun family	3	Yes						□			□
NN-disubstituted-2-aminoethyl-X (X: -OH, -SH, -Cl, -Br or -I)	VX family nerve gases	1–2	No						□			□
Phenyl-, alkyl- and cycloalkyl-glycollic acids	Some BZ family incapacitants	1	No						□			□
3- and 4-piperidyl compounds	Some BZ family incapacitants (not BZ)	1	No						□			□
Pinacolyl alcohol	Soman nerve gas	1	No						□			□

[a] The lists of countries and chemicals here may well be incomplete. The sources of information relied upon for details of the export controls were as follows: [301] for Australia (AUS), [353] for Canada (CAN), [89] for the United States (USA), [353, 64] for Japan (JAP), [11] for the European Communities (EC), [67] for the Netherlands (NL), [353] for Denmark (DK), [76] for the United Kingdom (GB) and [353] for Eire (IRL). The lists do not reflect the CW-related export controls which some governments (e.g., that of the USA) had in place already.
[b] Comprising Belgium, Denmark, Eire, France, FR Germany, Greece, Italy, Luxembourg, the Netherlands and the United Kingdom. Implementation of the EC Foreign Ministers' collective decision on export controls requires domestic legislation by each member state.
[c] By which is meant: are there realistic alternative ways of making the CW agent(s) concerned that do not go via the controlled chemical(s)?
[d] Exports to the USA are not controlled.

Ex: J P P Robinson, 'Chemical and biological warfare: developments in 1984', SIPRI Yearbook 1985 (London: Taylor & Francis, 1985) pp 159-219, at pp 174-75.

LITERATURE CITED

1 Report of the Ad Hoc Committee on Chemical Weapons to the Conference on Disarmament. Document CD/636* of 23 August 1985, at Article VI of Appendix I.

2 This paper combines the background paper ('Chemical-warfare weapons and relevant aspects of their manufacture') and the working paper ('Correlating categories of chemical with categories of control under the projected Chemical Weapons Convention') which the present author prepared for the conference.

3 J P P Robinson, 'Chemical warfare arms control: a framework for considering policy alternatives', SIPRI Chemical & Biological Warfare Studies (London: Taylor & Francis) no 2, October 1985.

4 See, further, SIPRI, The Problem of Chemical and Biological Warfare, 6 vols, Stockholm & New York, 1971-75, especially
(a) Vol 2, CB Weapons Today (1973)
(b) Vol 1, The Rise of CB Weapons (1971)

5 See the present author's 'Environmental effects of chemical and biological warfare' in War and Environment (Sweden, Ministry of Agriculture: Environmental Advisory Council, 1981), pp 73-117 at pp 93-95, which relies on newly opened British archives of documentation describing work on the subject at the time of World War II. See also: J Perera and A Thomas, '"This horrible natural experiment"', New Scientist, 18 April 1985, pp 34-35.

6 See, for example: Federal Republic of Germany, 'Working paper on the definition and classification of chemical warfare agents', Conference of the Committee on Disarmament document CCD/458 of 22 July 1975.

7 A Pittaway et al, 'Chemical Agent Production' (volume 2 of Inspection for Production, Transportation and Storage of Chemical and Biological Weapons), report by the Midwest Research Institute to the US Arms

Control & Disarmament Agency on contract no ACDA/ST-56, 18 August 1966, as cited in T L Ferguson, A R Hylton and C E Mumma, 'Analysis of Foreign Organophosphorus Chemical Technology' (volume 3 of Studies on the Technical Arms Control Aspects of Chemical and Biological Warfare), report by the Midwest Research Institute on contract no ACDA/ST-197, 13 November 1972, at pp 3-4.

8 A M Prentiss, Chemicals in War: A Treatise on Chemical Warfare, New York: McGraw-Hill, 1937.

9 The general propositions stated in this section are developed more fully in the present author's 'Chemical, biological and radiological warfare: Futures from the past', a submission to the (Palme) Independent Commission on Disarmament and Security Issues, September 1981. On the technical matters, see, further, the author's 'Chemical warfare: status of technological developments and deployments' in J Rotblat & S Hellman (eds), Nuclear Strategy and World Security (Annals of Pugwash 1984), London: Macmillan, 1985, pp 155-75.

10 References to the extensive literature on which this section relies are deliberately withheld, but are available from the author.

11 S Franke, Chemie der Kampfstoffe (Band 1 of Lehrbuch der Militärchemie), Berlin: Deutscher Militärverlag, 1967, at section 5.1.3.1.

12 D Oberdorfer, 'Chemical arms curbs are sought', Washington Post, 9 September 1985, p.A5.

13 This table is from a draft of the author's forthcoming CBW Data Book, which is currently in preparation for publication in the series SIPRI Chemical and Biological Warfare Studies. The multiple source documentation on which it is based will be cited there in full.

Appendix

CHEMICALS RELEVANT TO THE PROJECTED CHEMICAL WEAPONS CONVENTION THAT ARE ALREADY SUBJECT TO EXPORT CONTROLS IN THE UNITED STATES

Many specific chemicals are subject to export restrictions under US federal regulations. At least 65 of them are controlled because of their potential application in antipersonnel chemical warfare. Some are actual CW agents. Some are adjuvants. Some are precursors.

The Commodity Control List, which is the basis for the export controls administered by the Department of Commerce, identifies CW-relevant chemicals at Entry 4707B of Commodity Group 7 ('Chemicals, Metalloids, Petroleum Products and Related Materials'). This entry, which reflects the technology of the late 1950s, is reproduced in table A1 below. Elsewhere in the Commodity Control List are other chemicals that could also find CW-related applications. But only at Entry 5799D and Entry 6799G are such applications made explicit in any way: these are the entries that record the 'foreign policy controls' that were imposed on the eight chemicals listed in table A2 below following the verification last year of Iraq's use of chemical weapons against Iran.

The Commodity Control List thus identifies some 40-odd different chemicals as requiring special attention from the International Trade Administration of the Department of Commerce by virtue of their pertinence to antipersonnel chemical weapons. It is instructive to classify these chemicals according to the particular CW applications that can be envisaged for them. This is done in table A3. A bizarre picture is revealed.

Further export controls are mediated through the US Munitions List, which is the basis for the State Department's arms-trade controls. 'Toxicological agents', meaning chemical and biological warfare agents, are to be found in Category XIV. The text there indicates that the Munitions List is to apply to all 'chemical agents'. These are defined by the language reproduced in table A4 below. It will be seen that tear gases, herbicides and toxins (e.g., the capsaicin that is the active principle of the 'liquid pepper' listed) are all subsumed. Of the ten supertoxic and other lethal chemicals given as examples, two have civil applications: fluorine and chloropicrin. Neither phosgene nor hydrogen cyanide is referred to explictly. The Munitions List does not extend to precursors.

Table A1 Entry 4707B of the US Commodity Control List

Supplement No. 1 to § 399.1—CCL-74 Group 7 Commodity Control List

Export Control Commodity Number and Commodity Description	Unit	Validated License Required	GLV $ Value Limits T&V	Processing Code	Reason for Control
4707B (a) Chemicals, as follows:	Lb.	PQSTVWYZ	500[1]	MG	1

(1) Beta-diethylaminoethyl diphenylpropylacetate hydrochloride,
(2) 2-Chloro-10-(3-dimethylaminopropyl) phenothiazine,
(3) 2-Chlorophenothiazine,
(4) 2-Cyanoacetamide,
(5) 2-Dicyclohexylcarbodiimide,
(6) Diethylmethylphosphonite,
(7) Dihydrodibenzazepine,
(8) Dihydrodibenzocycloheptene,
(9) 10, 11-Dihydro-N, N-dimethyl-5H-dibenzo (a, d) cycloheptent delta 5, gamma-propylamine,
(10) 2-Diisopropylaminoethanol,
(11) Diisopropylcarbodiimide,
(12) 5-(3-(Dimethylamino)-2-methylpropyl)-10, 11-dihydro-5H-dibenz (b,f) azepine,
(13) Dimethylamino propylchloride hydrochloride,
(14) 5-(3-Dimethylaminopropyl)-10, 11-dihydro-5H-dibenz (b,f) azepine,
(15) 10-(3-Dimethylaminopropyl)-2-trifluorodimethylphenothiazine,
(16) Dimethyl hydrogen phosphite,
(17) Di-o-tolyl carbodiimide,
(18) Dissopropylaminoethylchloride hydrochloride,
(19) Ethylphosphonothioic dichloride,
(20) Ethylphosphonous dichloride,
(21) Lysergic acid diethylamine,
(22) Malononitrile,
(23) Methylbenzylate,
(24) Methyldichlorphosphine,
(25) Methylisonicotenate,
(26) Methylphosphonothioic dichloride,
(27) Methylphosphonous dichloride,
(28) Methylphosphonyldichloride,
(29) N, N-diethylethylenediamine (diethylaminoethylamine),
(30) Orthochlorobenzaldehyde,
(31) Phenothiazine,
(32) Piperidine carboxyl acid,
(33) n-Propylphosphonous dichloride,
(34) 3-Quinuclidinol,
(35) 3-Quinuclidinone,
(36) 2-Trifluoromethylphenothiazine, and
(37) 4-(3-(2-Trifluoromethyl) phenothiazine-10-yl) propyl)-1-piperazine ethanol.

(b) Synthetic organic agricultural chemicals, as follows:
(1) Alkyl aryl carbamates (including isopropyl N-phenylcarbamate and isopropyl N-(3-chlorophenyl)-carbamate),
(2) Aminochloropicolinic acid and its salts and esters,
(3) Bromoalkyl pyrimidines,
(4) Cacodylic acid,
(5) Chlorophenoxyacetic and chlorofluorophenoxyacetic acids and their salts and esters, and
(6) Herbicidal or antiplant preparations containing one or more of the above in amounts totaling, alone or in combination with another, 80 percent or more by weight.

[1] The GLV $ value limit for Country Group Q is $100, except for sub-entry (b), which is $500.

October 1, 1982 Export Administration Regulations

Ex: USA, Code of Federal Regulations, Title 15, Parts 368-399.

Table A2 Other chemicals in the Commodity Control List because of their chemical-warfare applications.

Commodity Description	**CCL Group 7 entry number**
Dimethyl methylphosphonate	5799D
Methyl phosphonyl difluoride	5799D
Dimethylamine hydrochloride	5799D
Potassium fluoride	6799D
Phosphorus oxychloride	6799D
Thiodiglycol	6799D
Ethylene chlorohydrin (chloroethanol)	6799D
Dimethylamine	6799D

Source: Federal Register 49: 13136 (3 Apr 84) and 49: (14 Sep 84).

Table A3 The Commodity Control List chemicals of Tables A1 and A2 classified according to their CW relevance

Class	Chemical referred to in the Commodity Description		Export Control Commodity Number
POTENTIAL CHEMICAL-WARFARE AGENTS (INCAPACITANTS)			
.	Tranquillizers		
. .	proadifen (SKF 525A)		4707B(a)(1)
. .	chlorpromazine		4707B(a)(2)
. .	triflupromazine		4707B(a)(15)
. .	fluphenazine		4707B(a)(37)
. .	amitriptyline		4707B(a)(9)
. .	trimipramine		4707B(a)(12)
. .	imipramine		4707B(a)(14)
.	Hallucinogen		
. .	LSD		4707B(a)(21)
ADJUVANTS FOR CHEMICAL-WARFARE AGENTS			
.	Storage-stabilisers for nerve gas		
. .	dicyclohexylcarbodiimide		4707B(a)(5)
. .	diisopropylcarbodiimide		4707B(a)(11)
. .	di-o-tolylcarbodiimide		4707B(a)(17)
PRECURSORS OF CHEMICAL-WARFARE AGENTS			
.	Precursors of nerve gases other than tabun-family ones		
. .	potassium fluoride		6799D
. .	$HOCH_2CH_2NPr^i_2$		4707B(a)(10)
. .	$[ClCH_2CH_2NHPr^i_2]Cl$		4707B(a)(18)
. .	$(MeO)_2POH$	dimethyl hydrogen phosphite	4707B(a)(16)
. .	$MePCl_2$	methyldichlorophosphine	4707B(a)(24)&(2)
. .	$EtPCl_2$	ethyldichlorophosphine	4707B(a)(20)
. .	$^nPrPCl_2$	n-propyldichlorophosphine	4707B(a)(33)
. .	$MeP(OEt)_2$	diethyl methylphosphonite	4707B(a)(6)
. .	$MeP(O)(OMe)_2$	dimethyl methylphosphorate	5799D
. .	$MeP(O)Cl_2$	methylphosphonic dichlcride	4707B(a)(28)
. .	$MeP(S)Cl_2$	methylphosphonothioic cichloride	4707B(a)(26)
. .	$EtP(S)Cl_2$	ethylphosphonothioic dichloride	4707B(a)(19)
. .	$MeP(O)F_2$	methylphosphonic diflucride	5799D

Table A3 (continued)

Class	Chemical referred to in the Commodity Description		Export Control Commodity Number
.	Precursors of tabun-family nerve gases		
. .	Me_2NH	dimethylamine	6799D
. .	Me_2NH_2Cl	dimethylammonium chloride	5799D
. .	$POCl_3$	phosphoryl chloride	6799D
.	Precursors of mustard gas		
. .	2-chloroethanol		6799D
. .	thiodiglycol		6799D
.	Precursors of tear gas (agent CS)		
. .	2-cyanoacetamide		4707B(a)(4)
. .	malononitrile		4707B(a)(22)
. .	2-chlorobenzaldehyde		4707B(a)(30)
.	Precursors of glycollate incapacitants		
. .	3-quinuclidinone		4707B(a)(35)
. .	3-quinuclidinol		4707B(a)(34)
. .	methyl benzilate		4707B(a)(23)
.	Precursors of other incapacitants		
. .	phenothiazine		4707B(a)(31)
. .	2-chlorophenothiazine		4707B(a)(3)
. .	2-trifluoromethylphenothiazine		4707B(a)(36)
. .	dihydrobenzazepine		4707B(a)(7)
. .	dihydrodibenzocycloheptene		4707B(a)(8)
. .	$[ClCH_2CH_2CH_2NHMe_2]Cl$		4707B(a)(13)
. .	$Et_2NCH_2CH_2NH_2$	2-diethylaminoethylamine	4707B(a)(29)
PRECURSORS OF NERVE-GAS ANTIDOTES			
. .	methyl isonicotinate		4707B(a)(25)
. .	either isonipecotic acid or pipecolic acid or nipecotic acid		4707B(a)(32)

Table A4 How CW agents are defined in the US Munitions List

§ 121.09 Chemical agents.

(a) (See Category XIV(a).) A chemical agent is a substance useful in war which by its ordinary and direct chemical action, produces a powerful physiological effect. The term "chemical agents" includes but is not limited to the following chemical compounds:

1. Lung irritants:
(i) Diphenylcyanoarsine (DC).
(ii) Fluorine (but not fluorene).
(iii) Trichloronitro methane (Chlorpicrin PS).
2. Vesicants:
(i) B Chlorvinyldichlorarsine (Lewisite, L).
(ii) Bisdichlorethyl sulphide (Mustard gas, HD or H).
(iii) Ethyldichloroarsine (ED).
(iv) Methyldichloroarsine (MD).
3. Lachrymators and tear gases:
(i) Brombenylcyanide (BBC).
(ii) Chloroacetophenone (CN).
(iii) Dibromodimethyl ether.
(iv) Dichlorodimethyl ether (ClCl).
(v) Ethyldibromoarsine.
(vi) Phenylcarbylamine chloride.
(vii) Tear gas solutions (CNB and CNS).
(viii) Tear gas orthochlorobenz almalononitrile (CS).
4. Sternutators and irritant smokes:
(i) Diphenylaminechloroarsine (Adamsite, DM).
(ii) Diphenylchlorarsine (BA).
(iii) Liquid pepper.
5. Nerve gases. These are toxic compounds which affect the nervous system, such as:
(i) Dimethylaminoethoxycyanophosphine oxide (GA).
(ii) Methylisopropoxyfluorophosphine oxide (GB).
(iii) Methylpinacolyloxyfluoriphosphine oxide (GD).
6. Antiplant chemicals:
(i) Butyl 2-chloro-4-fluorophenoxyacetate (LNF).

[Dept. Reg. 108.606, 34 FR 12029, July 17, 1969, as amended by Dept. Reg. 108.630, 35 FR 19994, Dec. 31, 1970; Dept. Reg. 108.646, 36 FR 20940, Nov. 2, 1971. Redesignated by Dept. Reg. 108.733, 42 FR 16617, Mar. 29, 1977]

CATEGORY XIV—TOXICOLOGICAL AGENTS AND EQUIPMENT; RADIOLOGICAL EQUIPMENT

(a) Chemical agents, including lung irritants, vesicants, lachrymators, and tear gases, sternutators and irritant smoke, and nerve gases and incapacitating agents (see § 121.08).

(b) Biological agents adapted for use in war to produce death or disablement in human beings or animals, or to damage crops and plants.

(c) Equipment for dissemination, detection, and identification of, and defense against the articles in paragraphs (a) and (b) of this category (see § 123.31).

(d) Nuclear radiation detection and measuring devices, except such devices as are in normal commercial use.

(e) Components, parts, accessories, attachments, and associated equipment specifically designed or modified for the articles in paragraphs (c) and (d) of this category.

Ex: USA, Department of Defense regulation DoD-5105.38-M, pages 2-11 and 2-13.

THE GENEVA TALKS ON CHEMICAL WEAPONS AND ATTITUDES DISPLAYED THERE TOWARDS THE CHEMICAL INDUSTRY

R Trapp

I Introduction

This background paper summarizes the history and the results of the negotiations held in Geneva within the international bodies that have been established there over recent decades: that is, the Ten- (later Eighteen-) Nation Disarmament Committee; the Conference of the Committee on Disarmament (CCD); the Committee on Disarmament (CD); and the current Conference on Disarmament (CD), devoted to the elaboration of an international treaty banning chemical weapons.

The main focus is on those questions which deal with the impact of such a treaty on the civilian chemical industry. A more general discussion covers the history of the negotiations and their present state. The paper is comprised of four parts. First, a historical survey of the Geneva talks is given. Second, a brief description of the present state of the negotiations and the major areas of agreement and disagreement are presented. Third, alternative approaches to the verification of the non-production of chemical weapons, with special emphasis on problems of the civilian chemical industry, are discussed. Finally, the proposed principles, procedures and areas of verification of the non-production of chemical weapons which originated from the working documents of the Geneva negotiation bodies are compiled in the appendices.

Before actually starting the historical survey, it seems advisable to recall that these negotiations are not the first attempt to ban chemical weapons, and that the envisaged Convention will not be the only international treaty in the field.

The fundamental agreement, which is widely regarded as a part of customary international law, is the 1925 Geneva Protocol which prohibits the use in war of asphyxiating, poisonous and other gases and of all bacteriological means of warfare. It is an agreed understanding today that the term 'gases' used in the Protocol applies to all warfare agents regardless of their aggregate state, and that 'bacteriological warfare' covers any means of biological warfare, not only bacteria. As at 31 December 1984 the Geneva Protocol had been ratified by 106 states, including the USSR (1928), the USA (1975), the UK (1930), France (1926) and China (1929).[1]

The scope of the Geneva Protocol is, however, limited in being a non-use agreement not affecting the possession of chemical weapons by its States Parties, and is further limited by the reservations of several states regarding the possibility of in-kind retaliation.[2] Moreover, the lack of any definition of chemical weapons (a term not actually used in the Protocol) has created diverging interpretations with respect to harassing agents (as in the case of the so-called 'tear gases') and anti-plant chemicals (in the understanding of the USA not covered by the Protocol). To overcome the limitation inherent in the scope of the Protocol negotiations, started already within the League of Nations, aimed at banning also the possession of chemical weapons. These negotiations failed to meet with success, however.

The 1925 Geneva Protocol was among those factors which played a role in deterring any large-scale use of chemical weapons during World War II. In the negotiations after World War II chemical and biological weapons were included in the talks on an international agreement on general and complete disarmament (GCD). They were covered in the relevant proposals made in preparation for the negotiations in Geneva at the end of the 1950s, as in: the UK address to the United Nations General Assembly dated 17 September 1959; the USSR Declaration on General and Complete Disarmament dated 19 September 1959; and the US Program for General and Complete Disarmament Under Effective International Control dated 27 June 1960.

II Brief history of the Geneva talks on chemical weapons

Chemical and biological weapons became the subject of the negotiations in the Ten- (later Eighteen-) Nation Disarmament Committee within the framework of negotiations towards general and complete disarmament. These weapons had been covered in the joint Soviet-US statement on agreed principles for disarmament negotiations (20 September 1961) as well as in the subsequent proposals of the USSR, the USA and the UK. The central aims of these proposals for general and complete disarmament with respect to chemical (and biological) weapons were: to eliminate the stockpiles of these weapons (and of nuclear weapons); to stop their production; and to eliminate from military arsenals the means of delivery of these weapons.

There were, however, diverging approaches in major areas such as the scope of measures to be taken under a GCD treaty, time schedule for these activities and verification.

In the late 1960s the issue of chemical and biological weapons received greater attention in public discussions as well as in negotiations both within the United Nations and in Geneva, primarily because of the mass-scale use of irritants and anti-plant chemicals by US troops in Viet Nam. This was reflected in a 1966 UN General Assembly resolution which called, _inter alia_, for the strict observance of the Geneva Protocol.

In 1968 the question of prohibiting chemical and biological weapons entered the provisional agenda of the Eighteen-Nation Disarmament Committee, as a separate item. In August of the same year the British delegation presented, for the first time, a proposal for dealing separately with chemical and biological weapons, arguing that the military significance of both differs considerably, as do the verification requirements. In the opinion of the British delegation, it was much easier to reach agreement on banning biological weapons than on banning chemical weapons. Views on this approach were split and a controversial discussion took place over the following years. In 1969 the Secretary-General of the United Nations submitted a report to the General Assembly in which chemical and biological weapons were treated together in an expert analysis of the consequences of the use of these weapons. They were also discussed together in a comparable

WHO expert study on the same subject published in the same year.

In his report, the Secretary-General renewed the appeal to all states to accede to the Geneva Protocol, made a clear affirmation that the Protocol´s prohibition relates to all means of chemical and biological warfare, and called upon states to reach agreement to halt the development, production and stockpiling of all chemical and biological weapons and to eliminate the existing stockpiles of these weapons. Harassing agents (´tear gases´) and anti-plant agents were included in the Secretary-General´s statement. The report was welcomed by the CCD, although dissenting views were expressed on specific questions.

In 1969 President Richard Nixon announced a unilateral US moratorium on the possession of biological weapons (one year later toxin weapons were also included in this moratorium) and the production of further lethal chemical weapons. Simultaneously, the US policy of no-first-use of lethal chemical weapons was reaffirmed and extended to incapacitating agents. In the same year, the socialist countries presented their draft Convention for a ban on chemical and biological weapons in the CCD. The UK delegation developed its position of a separate treatment of chemical and biological weapons by submitting a draft Convention prohibiting (micro)biological weapons.

The controversy about a separate approach ended in 1971 when the socialist states submitted to the CCD a draft Convention banning the development, production and stockpiling of biological weapons, and on their destruction, thereby accepting generally the British position. This draft was revised taking into account the views of the Western countries and of the non-aligned states, and was presented again on 30 March 1971, supported by an identical US proposal. After further amendment the CCD submitted the draft BW (Biological Weapons) Convention to the UN General Assembly at the end of 1971. It entered into force in 1975 and currently binds 101 States Parties including the USSR (1975), the USA (1975), the UK (1975), France (1984) and China (1984).[3]

In the area of chemical weapons, the negotiations in Geneva continued, and these weapons received greater attention during subsequent years.

Among the group of Western states, a very detailed discussion of specific items began emphasizing primarily the definition of`chemical weapons, the relation of a future chemical weapons ban to the civilian chemical industry and the problems of verification. In 1972 the socialist countries proposed a draft Chemical Weapons Convention with wording which was identical to that of the BW Convention. This approach was, however, rejected by the group of Western states, arguing that it lacked the essential verification requirements. The draft Convention envisaged, as in the BW Convention, a complaint procedure with the UN Security Council.

The following years were characterized by more and more detailed discussion on matters such as toxicity determination, and the definition of chemical weapons; the scope of the future Convention; verification requirements, identification and quantification of chemical-warfare agents; problems associated with the destruction of chemical weapons; and medical protection against chemical-warfare agents.

In the discussions regarding the scope of a future Convention, two major areas may be identified. The first was the question of whether the initial step should be either a comprehensive ban on all chemical weapons or a limited ban on the most dangerous chemical weapons (the lethal agents). An important part of this discussion was the question of the inclusion of harassing and anti-plant agents under a comprehensive approach. The second area related to the activities prohibited or permitted under the Convention (research and development; large-scale production vs. permitted small-scale production of chemical agents for purposes not covered by the potential Convention; destruction of CW stocks and their former production facilities--including filling facilities; transfer of chemical weapons; and so on).

These discussions must be seen in close relation to those dealing with the definition of chemical weapons (application of a ´general purpose criterion´ vs. other criteria such as toxicity; clarification of relevant terms such as ´dual-purpose agent´, ´precursor´, ´key precursor´, and the like).

The central point of controversy throughout the discussions was verification. The group of Western states held the view that the disarmament of chemical weapons, which had already been included in military arsenals, required international-verification procedures including obligatory on-site inspections. The group of socialist countries meanwhile favoured national-verification procedures, arguing that perfect control in the field of chemical weapons is both impossible and unnecessary.

In April 1973, 10 non-aligned states submitted a working paper on the prohibition of the development, production and stockpiling of chemical weapons and on their destruction.[4] This proposal was not formally labelled as a draft treaty, but was nevertheless a comprehensive presentation of the non-aligned view on this matter. The document suggested general provisions for the proposed Convention, the scope of the ban, verification means as well as systems of control and the procedure for complaint.

Under its general provisions, the document emphasized the obligations assumed by the Parties to the 1925 Geneva Protocol. Simultaneously, it stressed that the Convention should be implemented in a way designed to avoid hampering the research, development, production, possession, transfer and application of chemical agents for peaceful purposes; and to refrain from hindering the economic and technological development of the States Parties. The treaty should be designed so as to facilitate the fullest possible exchange of chemical agents, equipment and material, as well as of scientific and technological information for the peaceful use of chemical agents.

With respect to the scope of the Convention the working document started with a general-purpose criterion but recognized the need for additional criteria.

For verification, a combination of national and international measures was envisaged. The minimum basic elements of a verification regime were indicated as follows: self-control of the States Parties; national means of verification ('national technical means'); and international measures for verification. Provision was made for the establishment of an international-verification organ (International Control Organization). National-verification measures (and related activities) mentioned were: declarations regarding national activities related to the production and development of chemical weapons, particularly concerning their destruction; measures aimed at the implementation of the Convention; organization of a national system of control, including a national control body; and information transfer to the international body.

International means of verification were to be comprised of the collection, analysis and circulation of relevant data, and assistance to States Parties in the development of their elements of self-control as well as their national means of verification. The verification regime encompassed all activities related to the development, production and stockpiling of chemical-warfare agents. As a non-recurrent measure, special provision was made for international inspections in order to verify the destruction of stockpiles. Finally, a complaint procedure with the UN Security Council was proposed.

On 30 April 1974 the Japanese delegation presented its government's draft Convention on the prohibition of the development, production and stockpiling of chemical weapons, and on their destruction.[5] This draft Convention had a comprehensive approach in prohibiting the development, production and stockpiling of chemical weapons. Also prohibited were the acquisition or retention of chemical agents of such types and in such quantities as could not be justified for protective or other peaceful purposes, as well as weapons, equipment and those means of delivery designed to use chemical agents for hostile purposes or in armed conflicts. There was no further definition of chemical weapons, but a general-purpose criterion was applied.

The draft called for the establishment of an International Verification Agency (IVA) having, *inter alia*, the following functions: to analyse and evaluate periodic reports and statistical and other documents submitted by each State Party; to request explanations and conduct enquiries; to conduct inspections; and to consult and co-operate with national authorities (organs). Under article IX the draft Convention set regulations for international inspections to be carried out by the IVA when states failed to provide adequate explanations after enquiries had been made. States were therefore urged to make every effort to accept, as soon as possible, such on-site inspections, and to give adequate reasons if inspections were deemed unacceptable. The Japanese draft Convention was obviously intended to serve as a compromise among the diverging approaches within the CCD, but it failed to receive support as a broadly accepted basis for further negotiations.

In 1976 the United Kingdom submitted its draft Convention on a comprehensive ban on chemical weapons to the CCD.[6] It covered lethal as well as other toxic chemicals ('of a nature and intended primarily to cause long-term physiological harm to human beings'), again applying the general-purpose criterion, as well as urging the prohibition of munitions, equipment or systems designed to deliver such agents for hostile purposes or in armed conflicts. The draft Convention called for a number of declarations and notifications to be made at the time of signing or acceding to the Convention, including declarations on: the possession of chemical weapons, their types and quantities; the existence of production facilities for chemical weapons; information regarding all production facilities on a State Party's territory capable of producing the agents, munitions, equipment or systems; information on types and quantities of agents produced for protective or other peaceful purposes--including production locations and factories involved; and information on the national organization in charge of collecting the information described above.

The draft Convention included the proposal for establishing a Consultative Committee of the States Parties to: analyse and evaluate the periodic reports and other information provided by the States Parties; call for supplementary information when deemed necessary; request information and conduct inquiries if asked to do so by States Parties; verify the destruction of stockpiles of chemical weapons and conduct other inspections; notify and inform States Parties following verification procedures; and consult and co-operate with the national organizations or authorities of the States Parties.

In the field of verification, the UK draft envisaged: mandatory inspections of the former chemical-weapons production facilities within 6 months after entry into force of the Convention; sealing of these facilities and periodic inspections thereafter; a number of on-site inspections (up to 10) annually per State Party in chemical-industry facilities declared to be capable of producing the prohibited agents, and the other declared facilities (see above). (Factories should be examined to ensure that no production of the specified agents was occurring.) And inspections during the destruction process of chemical agents as the Consultative Committee agreed to be necessary for the task of verification.

Article X of the British draft included obligations for the States Parties to accept international on-site inspections which the Consultative Committee deemed necessary. The British draft Convention was supported by the group of Western states but rejected by the group of socialist countries mainly due to its emphasis on international on-site inspections.

In 1979 the CCD was re-established, under the name Committee on Disarmament (CD). Negotiations towards an international treaty banning chemical weapons were given a high priority on the Committee´s agenda. This aim was furthered by joint Soviet-US efforts, based on an agreement which had been reached in 1974 to co-operate in a ´joint initiative´ towards the elimination of, what was referred to at that time as, ´the most dangerous, lethal means of chemical warfare´.

The first progress report of these bilateral talks was presented jointly in 1979[7] and dealt with all chemical weapons, not only lethal ones. Both States agreed on a comprehensive approach to the scope of the proposed Convention. States Parties would be obliged by the Convention: (a) not to develop, produce, stockpile, acquire, possess or retain chemicals for chemical-weapons purposes; (b) not to transfer to anyone any means of chemical warfare and not to assist, encourage or induce others to carry out prohibited activities; (c) to destroy all stocks of chemical weapons; and (d) to destroy or dismantle all former production facilities for chemical weapons, and munition-filling facilities for chemical weapons.

For the definition of chemical weapons a general-purpose criterion was chosen as a basis, supplemented by toxicity criteria placing the chemicals in the following categories:

- ´supertoxic lethal chemicals´, their toxicity being higher than a certain limit (LCt_{50} = 2000 mg m^{-3} min^{-1}, LD_{50} = 0.5 mg kg^{-1} subcutaneously);
- ´other lethal chemicals´ being less toxic than supertoxic lethal chemicals, but more toxic than given by the following limit: LCt_{50}

(inhalant) = 20 000 mg m^{-3} min^{-1}, LD_{50} = 10 mg kg^{-1} (subcutaneously); and

-´other harmful chemicals´ being less toxic than other lethal chemicals.

It was made clear that this grouping related to different verification requirements.

The two Parties agreed on the necessity of the submission of initial declarations by the future States Parties regarding the existence or non-existence of stocks of chemical weapons as well as of production or munition-filling facilities for chemical weapons. Simultaneously, plans for their destruction or dismantling should be declared. The USSR and the USA agreed also, in general terms, on the duration of the disarmament measures (10 years was envisaged for the entire stock-destruction operations, destruction of initially sealed production facilities to begin not later than 8 years after entry into force of the Convention and to be completed not later than 10 years thereafter). The Convention also foresaw the need for proper protection of the people involved in destruction operations, and of the environment as well.

For verification, both sides agreed on the necessity of ´adequate verification´ composed of national and international procedures and means. A Consultative Committee should be established to handle all issues of co-operation, consultation, information and data exchange, and verification. For technical reasons the establishment of a permanent secretariat was envisaged.

In an assessment in the 1980 SIPRI Yearbook,[8] it was concluded that: ´While the two sides agree that in the case of civilian chemical industry routine checks would be both impractical and undesirable, and that optional inspection, "by challenge", would be enough, they differ as regards the need for mandatory international on-site inspection in other cases´. The US delegation made clear that it insisted upon mandatory international on-site inspections for: the destruction of declared stocks of chemical weapons; the moth-balling and eventual destruction of chemical-weapons production and filling facilities; and checks of the facility(ies) permitted to produce chemicals (otherwise prohibited by the Convention) for permitted purposes.

The bilateral talks were continued and results presented jointly in July 1980,[9] but the above-mentioned differences remained at the time. In the same year, the US delegation indicated its unwillingness to continue further in these bilateral discussions using the ´Yellow Rain´ allegations as an official argument.

In 1980 the *Ad Hoc* Working Group on Chemical Weapons was established formally by the Committee on Disarmament. This subsidiary body was later reorganized as the *Ad Hoc* Committee on Chemical Weapons and in recent years has formed the major multilateral negotiating structure. Its role has expanded over the years, now being a clear mandate for negotiating a draft text of the future CW Convention, with the exception of its final drafting. Basic documents under consideration by this multilateral body are the bilateral Soviet-US documents, and more recently working documents and drafts such as:

- CD/294 (USSR, 21 July 1982) ´Basic provisions of a convention on the prohibition of the development, production and stockpiling of chemical weapons and on their destruction´;
- CD/343 (USA, 10 February 1983) ´United States detailed views on the content of a future chemical weapons ban´;
- CD/443 (PR China, 5 March 1984) ´Proposal for major elements of a future convention on the complete prohibition and total destruction of chemical weapons´; and
- CD/500 (USA, 18 April 1984) ´Draft convention on the prohibition of chemical weapons´.

Elements of earlier drafts (as, for example, the draft CW Conventions of the group of socialist countries, the UK and Japan) and working documents have been also under consideration by the subsidiary group as well as a number of more specific working documents from several countries. The *Ad Hoc* Working Group (Working Committee) presented its results in several reports to the CD, labelled CD/131.Rev.1, CD/179+Add.1 (a chairman´s report, not an agreed document of the group), CD/281.Rev.1, CD/334, CD/342, CD/429 and CD/539.

The work was performed on several levels including: whole-group (committee) sessions, discussions in sub-groups, sessions with technical experts (formally labelled 'Chairman's Consultations with Technical Experts'), meetings of 'contact groups' and other, even more informal, bodies. These multilateral efforts were accompanied by bilateral and multilateral consultations.

Major issues in these discussions were: the scope of the future Convention and obligations to which States Parties would be bound; the definition of chemical weapons and of terms used in the Convention to characterize chemicals, activities and the like; declarations to be prepared by the future States Parties, how these declarations should be scheduled, how detailed they should be and so on; verification requirements (both from the technical and political points of view), and principles for verification, co-operation and so forth; the role and function of the proposed Consultative Committee and its proposed subsidiary organs; and the preamble, entry into force, withdrawal clause and other legal questions. The multilateral negotiations proceeded to preliminary drafting beginning in 1984. The present state of these negotiations is outlined in the following section.

III The present state of the negotiations

The discussion which follows is structured according to some major areas of the negotiations.

Scope

The agreed document of the *Ad Hoc* Working Committee of the CD on chemical weapons of 1984[10] defines a comprehensive scope for the CW Convention covering prohibitions on: the development, production, or otherwise acquiring, stockpiling or retaining of chemical weapons, or their transfer, directly or indirectly, to anyone; the assisting, encouraging or inducing, in any way, anyone to engage in activities prohibited to Parties to the Convention; and the use of chemical weapons (disagreement remains as to the wording of this prohibition).

The proposed Convention also obliges States Parties to: destroy their chemical weapons (there still remains some disagreement as to the possibility of the conversion of parts of chemical weapons for peaceful purposes, and as to the detailed wording); and destroy or dismantle chemical-weapons production facilities (again some disagreement on the detailed wording). Further proposals cover preparatory activities for the use of chemical weapons (as, for example, in the concept of ´chemical warfare capability´), but there is certain disagreement on this approach.

Besides the general agreement on certain basic questions related to the scope of the Convention, some sensitive problems are not finally solved. The inclusion or non-inclusion of herbicides and riot-control chemicals (such as ´tear gases´) in the prohibition is an example of such a problem. The US delegation considers the use of riot-control agents (besides their application for domestic law enforcement which is more or less generally accepted) for certain military purposes to be allowed under the Convention.

This is an indication of the broader field of controversy related to the coverage by the treaty of so-called ´dual-purpose agents´, especially in the cases of precursors and key precursors. Another complicated area is the coverage of toxins by the CW Convention. These substances are already prohibited by the Biological Weapons Convention, but are chemical in nature. Potential problems may arise with regard to differing verification regimes under both Conventions.

Declarations

It is commonly agreed that States Parties will be obliged to make certain declarations on: their possession or non-possession of chemical weapons; their possession or non-possession of--actual as well as former, but possibly limited by a certain date--production facilities for chemical weapons, including munition-filling facilities; and their plans for the destruction or dismantling of both chemical weapons and chemical-weapons production and filling facilities.

Controversy remains, however, as to the specific content of

certain declarations, and as to their timing. The USA insists that detailed declarations be made 30 days after either entry into force or accession to the Convention, including the disclosure of the location of both stockpile areas for chemical weapons and their production facilities, in order to prevent their illicit use during the period between declaration and destruction. Declarations of chemical-weapons possession should therefore be in detailed terms including types of chemical and ammunition, and their numbers.

The Soviet Union opposes this position with respect to chemical-weapons depots, favouring a general declaration of possession or non-possession and of the aggregate quantity of agents after entry into force or accession to the treaty, and a step-by-step detailed declaration of specific items immediately before their destruction (that is, in the respective plans for destruction) at declared destruction facilities. France has also stated that it prefers a grouping of the stockpiles at destruction sites and detailed declaration of them simultaneous with the plans for destruction. This approach was also supported by the Netherlands. The basic argument is that immediate declaration of locations and structures of chemical-weapons stocks may render them vulnerable to attack in the event of war.

Other areas of disagreement relate to the complicated problem of a co-ordinated timetable for the 'balanced' destruction of stockpiles of chemical weapons guaranteeing the security interests of all States Parties.

Definitions

For reasons of clarity, the CW Convention will contain certain essential definitions of sensitive terms used in the treaty. Examples are terms such as 'chemical weapons', 'precursor', 'key precursor', 'dual-purpose agents', and 'permitted activities'.

General agreement seems to have been reached as to the application of a general-purpose criterion. Thereby, the existence of 'dual-purpose agents' with both military and civilian application is acknowledged, the production of which will not be affected by the Convention if done for legitimate civilian purposes. Their production, possession, stockpiling and so on will be prohibited only when carried out as preparations for chemical-warfare activities. This, however, leads to a complicated situation with respect to verification (which will be discussed later) and a need for clear definitions. These definitions may then form a basis for listing chemicals for different levels of control. Other necessary definitions relate to terms such as ´chemical weapons production facility´.

In defining ´chemical weapons´ the current general approach is to include: toxic chemicals and their precursors, including the components of binary chemical weapons, except for those intended for permitted purposes, as long as the types and quantities involved are consistent with these permitted purposes; munitions and devices specifically designed to cause death or other harm through the toxic properties of the toxic chemicals; and any equipment specifically designed for use directly in connection with the employment of such munitions or devices.

The term ´toxic chemical´ is further differentiated by the application of toxicity criteria, as already discussed with respect to the joint Soviet-US documents CD/48 and CD/112. The categories are ´supertoxic lethal chemicals´, ´other lethal chemicals´ and ´other harmful chemicals´.

The term ´precursor´ is understood to cover chemical reagents which take part during the production of a given toxic chemical.

It is commonly accepted that a definition of chemical weapons should also cover binary chemical weapons. However, their existence complicates agreement on the essential term ´key precursor´--those precursors which pose a significant risk to the objectives of the Convention. One possible solution for this problem was indicated by the application of a term ´key component´ in recent discussions.

After long discussions, the major suggestions for the definition and criteria of the term ´key precursor´ were the following: it should play an important role in determining the toxic properties of the final chemical agent; it should be used in a final (even <u>the</u> final) state of the production of the toxic chemical only; it should (or should not) also cover the chemicals applied in binary systems; and it should not be used in any significant quantity for peaceful purposes.

There still remains a broad range of disagreement, or at least divergent approaches, with respect to these criteria and definitions, being indicated also by the reluctance of certain states to involve themselves deeply in the drawing up of certain lists of chemicals for verification purposes.

Verification

Verification remains by far the most controversial area of the negotiations.

There is some common agreement on certain fundamental principles for the final verification regime, such as:

(a) the necessity to combine national procedures of verification, national-technical means and international verification procedures;
(b) the requirement for an international Consultative Committee of the States Parties which, *inter alia*, will be responsible for international-verification procedures, systematic as well as on-challenge international on-site inspections, information exchange, co-operation with the corresponding national authorities of the States Parties, the international supervision of the implementation of the Convention, and the like;
(c) the necessity of a possibility for *ad hoc* (on-challenge) international-verification procedures (including necessary on-site inspections) and a complaint procedure with the Security Council of the United Nations;
(d) the performance of certain mandatory systematic international on-site inspections in well-defined cases: during the destruction of declared stockpiles of chemical weapons, during dismantling or destruction of former production facilities, and for the verification of a permitted small-scale production of supertoxic lethal chemicals and their key precursors in a single specialized facility per State Party having a capacity which is limited in a way that the annual stockpiling of the State Party does not exceed one metric tonne (on a regular periodic basis); and
(e) the necessity to establish the respective national capabilities (for example, a national agency or organ), to guarantee the implementation of

the Convention, to supervise this process on behalf of the respective governments, and to co-operate with the international Consultative Committee (including the preparation of reports and statistics to be submitted to the Consultative Committee, co-operation during international inspections and so on).

However, several major areas of disagreement remain in the field of verification, the most important being the problem of the mandatory vs. voluntary character of on-challenge on-site inspections, especially (but not exclusively) as exemplified by the most recent US proposal contained in the US draft CW Convention on ´open invitations´ for special inspections. This concept will be briefly discussed later.

Another question, especially relevant with regard to the involvement of the civilian chemical industry, is the approach of regular (systematic) random inspections in certain civilian enterprises vs. the approach of on-challenge inspections solely.

The concept of randomly selected international inspections in civilian chemical enterprises was developed by states such as FR Germany, the UK and the Netherlands. It will be given some consideration in section IV. The concept is supported within the group of Western states and by certain non-aligned countries, but the group of socialist countries and perhaps some other non-aligned countries take the view that such regular (mandatory) inspections should be limited to national authorities and international inspections should be restricted to on-challenge procedures. They argue that regular international inspections would be both an unacceptable burden for the states and ineffective in detecting clandestine chemical-weapons production, especially in the light of existing binary technology. They also point to the limited resources a future Consultative Committee will have to cope with the numerous international-inspection tasks required. A comprehensive international inspection of the relevant parts of the civilian chemical industry is said to be both impossible and inadvisable, and the random-selection procedure which is aimed at reducing the burden of such efforts is not accepted by this state group.

By far the most controversial concept regarding verification of a CW Convention is the above-mentioned ´open invitation´ enshrined in the US draft. In submitting the US draft CW Convention in April 1984 (document CD/500), Vice-President George Bush underlined that this concept is regarded as basic to the draft. In article X the proposal is made that each State Party must consent at 24-hours notice to a special inspection ordered by one member of the ´fact-finding panel´, a subsidiary body of the Consultative Committee, for one of the sites committed to systematic international on-site inspections or, more importantly, to <u>any</u> location or facility owned or controlled by the respective government of a Party, including military and civilian (that is, industrial) facilities. It was later made clear that controlled´ meant controlled by ´contract, other obligations or regulatory requirements´. This approach was categorically rejected by the USSR pointing out that it would result in the potential disclosure of political, economic, scientific, military, commercial and other secrets of a Party. The concept was also characterized as discriminating against countries with a state-owned or partly nationalized economy.

As a matter of fact, the open-invitation proposal has no precedent in recent disarmament agreements and seems to envisage absolute guarantees under the Convention. This aim has to be questioned in principle, taking the present political and economic situation in the world into account.

The concept of ´open invitations´ received poor, if any support, but was stressed again by the US delegation during the 1985 spring session.[11] One consideration within the CD negotiations was that, should this approach be implemented, problems could arise from unfounded recriminations thereby creating politically unacceptable burdens. Nor was the degree of absolute guarantees demanded by the USA deemed necessary, considering the limited scope of the future CW Convention (which covers only one area in military arsenals), and considering the actual military significance of chemical weapons as compared with other weapons (especially with respect to nuclear systems). For a long period it has been commonly acknowledged that absolute control within a Chemical Weapons Convention is both unnecessary and unachievable, and that insistence on such guarantees is politically cumbersome.

The 1985 SIPRI Yearbook concluded that: ´Under a verification system, which has already been broadly accepted as comprising both human supervision and automatic monitoring instruments, large-scale violations involving militarily relevant quantities of chemical weapons in readily useable form could not remain undetected, even without "special inspections" proposed by the USA´.[12] A certain element of uncertainty has to be accepted by the future States Parties of any disarmament agreement given the present world situation.

In conclusion, it must be underlined that in the field of verification major areas of disagreement remain which, at least under present prospects, may ultimately prove to be the most significant block to the rapid completion of a Chemical Weapons Convention.

IV Approaches for the verification of non-production of chemical weapons in the civilian chemical industry

Part of the inherent complexity of this specific issue has already been addressed in J.P. Perry Robinson´s background paper. There is no definite borderline between civilian and military chemicals, indeed some of the older chemical-warfare agents are exploited on a large scale as civilian intermediates and are produced in significant quantities in any developed chemical industry throughout the world. Typical examples are hydrogen cyanide, chlorine and phosgene. Moreover, there are certain industrial chemicals and pharmaceuticals which carry an even higher mammalian toxicity than those just mentioned. Finally, there are close relations both in chemistry and in toxicology between potent nerve agents and certain organophosphorus pesticides produced in huge quantities world-wide.

This complicated situation, which creates major difficulties with respect to the verification of a treaty banning chemical weapons, had been foreseen from the outset of the CW negotiations in Geneva. The conclusions drawn differ considerably, however. On the one side, the USA has argued that any simple statistical monitoring of certain sensitive material flows in national economies would not produce conclusive results and would alone require the surveillance of some 90 different chemicals in the phosphororganic chemical industry of a State Party.[13] The US delegation therefore emphasized international control mechanisms based on on-site inspections. On the other side, the USSR has acknowledged the inherent complexity but concluded that only national verification seems to be possible, advisable and necessary.[14] The USSR has accepted, however, the possibility of complaint procedures with the UN Security Council.

Somewhere in between these positions were early suggestions proposed by certain states (such as Japan, Yugoslavia, Sweden and Italy) emphasizing national-statistical surveillance of the material flows of some sensitive chemicals, mainly organophosphorus intermediates (for details see appendix III). Such surveys were to trace the production, export, import, stockpiling and use of certain chemicals within a nation. Regular reports to an international body for evaluation were foreseen. The concept was later taken up by the group of socialist countries[15] but, at least implicitly, rejected by the USA which underlined the difficulties of such evaluations for detecting clandestine CW production.[16] This approach was also criticized by France.[17]

Nonetheless, the concept of tracing the flow of certain sensitive material within the national economies of the States Parties has survived and forms one of the basic approaches to verification today, as a supplement to other verification techniques. However, disagreement exists as to which chemicals should be traced, and by whom.

National organizations are favoured widely as concerns those ˊdual-purpose agentsˊ which have a legitimate civilian application. International tracing based on systematic on-side inspections is commonly regarded as necessary with respect to control of the permitted production of supertoxic lethal agents and their key precursors in a single, specialized, small-scale production facility per State Party.

However, national surveillance and reporting to an international body (the Consultative Committee), even when combined with a complaint procedure enabling on-challenge international investigations including on-site inspections if necessary, are seen by certain states to be insufficient. During the long period of negotiations the group of Western states favoured the inclusion of at least some international-verification procedures (including on-site inspections) for other than the presently agreed cases.

Two approaches, which seem to be considered supplementary, were near-site inspections (this proposal was developed originally in discussions within the Pugwash Chemical Warfare Study Group) and on-site inspections on either a quota basis or, as currently proposed, a random-selection basis. The concept of near-site inspections was promoted by investigations in the Netherlands which have been submitted to the CCD and the CD. That study covered nerve agents of the G- and V-type including binary weapons and related to the 'fingerprint' concept of detecting methyl-phosphorus chemicals based on the downstream sampling and analysis of water (waste water, stream water) at chemical factories. It was argued that methyl-phosphorus chemicals are not used in pesticide production (because of their mammalian toxicity) and are rarely present, if at all, in nature; their detection should therefore be an indication of possible clandestine nerve-agent production.

It is worth remarking here that certain states have also proposed banning, in order to ease verification, the entire class of methyl-phosphorus chemicals from (civilian) production. A firm proposal on this line was tabled by the USSR delegation in Geneva.[18] Other delegations suggested a prohibition of these chemicals over a certain period, starting with the entry into force of the Convention. However, it was pointed out that certain areas of peaceful application of methyl-phosphorus chemicals exist (as in the case of some flame retardants), and that future fields of application may appear.[19] The potential production of methyldichlorophosfine as a precursor for a new herbicide has been indicated by the FRG delegation.[20]

The major controversy currently with regard to this area of verification seems to be the acceptance or non-acceptance of some regular international on-site inspections in civilian chemical enterprises. The group of Western states emphasizes the need for such inspections, whereas the group of socialist states consistently favours national verification in this area, combined with reporting and possible on-challenge international mechanisms.

There is presently one example of a regular international verification of the non-production of chemical weapons within the civilian chemical industry, that is, the WEU (Western European Union) controls of the chemical industry in FR Germany. This procedure has been described in working documents to illustrate to the CD delegations its design and efficiency.[20,21] The procedure is a step-by-step approach starting from relatively non-intrusive observations tracing special safety precautions in the facility and proceeding to more intrusive levels in suspicious cases. The concept has been applied in several CD documents and developed further as a proposed international CW Convention by: introducing a random-selection procedure to minimize the verification efforts (for example, by introducing a lottery system); and including inspection of certain chemical, biochemical and physico-chemical analyses in process streams, intermediate stocks, raw material stocks, and final products to cope with the problem of binary technology.[21,22]

Originally, the concept related to phosphororganic chemicals, only. The UK delegation applied it in a more general way when developing its proposal for the verification of non-production of chemical weapons, in general, in the civilian economy of the States Parties. Based on discussions regarding the definition and classification of chemicals, precursors and key precursors, the UK proposal[20] groups chemicals in different categories based on their risk both in terms of toxicity and of the specific risk they pose to the CW Convention, as follows:

High-risk chemical agents, covering the supertoxic lethal agents production of which is prohibited except in the specialized small-scale facility in each State Party, under international supervision. Chemicals concerned here were the phosphororganic nerve agents, sulphur and nitrogen mustards, lewisite, and glycollate incapacitants. Should such chemicals be produced for civilian purposes, regular reporting including a description and justification of the civilian use and random on-site inspection would be required.

High-risk precursors, covering certain sensitive intermediates for chemical agents (for details see appendix III) and being supervised as for high-risk chemical agents.

Medium-risk chemicals, coping with the category of 'dual-purpose agents'. All civilian production should be reported (regular reporting including information exchange and data exchange on production statistics).

Medium-risk precursors, covering the less important (for the determination of the toxic properties of the final product) or remote (regarding the production stage) intermediates of CW chemicals. Verification was conceived as for medium-risk chemicals.

The concept of the UK delegation is thus a combination of national reporting and random international inspections. The latter had been described in an earlier UK working document,[23] objectives being: to ensure that the quantities of a particular substance being produced at the facility inspected are compatible with the declared use; that any stockpiling is carried out in a manner and quantity compatible with the declared civilian use; and that the production facilities have not been modified in a way to enable the production of chemical-warfare agents.

On-site inspections were to consist of: examinations of the production lines for the facility concerned; visual observations on-site (inside and outside the production facility, detection of stockpiling areas and facilities which seem to be unnecessary, detection of munition filling facilities, detection of over-specialized safety equipment and the like); and engineering inspections to ensure that the production line is compatible with the production of the declared substance.

For the statistical accounting, a declaration of all facilities capable of producing the high- and medium-risk chemicals as well as precursors was envisaged in UK document CD/514 covering:

- the chemical name and formula of the chemical produced;
- name of the company/organization operating the plant;
- location of the production site (postal address, geographical co-ordinates);
- whether the chemical is solely for domestic use or also for export;
- the states to which the chemical is exported, if any export;
- whether the chemical is produced in dedicated plant or by batch process;
- if in a dedicated plant, the maximum annual capacity (tonnes per annum);
- if in batch, the production weight (in tonnes) of the last year;
- whether the chemical is stored at the site; if yes, the maximum storage capacity; and
- whether the chemical is used on stream.

These declarations should cover any production exceeding one metric tonne per annum.

The UK concept was further developed in 1985 with the presentation of two working documents regarding inspection procedures[24] and the role of the involved international organs[25] regarding the random inspection of facilities capable of producing high-risk chemicals/precursors.

In defining the framework of the inspections, the experience of the International Atomic Energy Agency (IAEA) and the Treaty on the Non-Proliferation of Nuclear Weapons (NPT) safeguarding regime was taken as a starting point. It was acknowledged that the situation is much more complex for chemicals where no single-element tracing can be successfully applied. Therefore, the inspectors for the random inspections should be authorized to: examine the relevant factory records; make independent measurements at agreed points; check the measurement and control equipment used at the facility; observe facility measurements, sampling and calibration; and take duplicate or additional samples and transfer them, for subsequent analysis at authorized laboratories.

The UK delegation stressed that the proposed procedure was aimed at reducing as much as possible any burden on the day-by-day performance at the inspected civilian facility. The procedure was described so as not to lead to any unnecessary disclosure of commercial secrets or any other unrequired encumbrance on the operation and further development of the facility.

With respect to data exchange, the UK delegation offered a differentiated approach of annual reporting, placing the emphasis for detailed information on the high-risk chemicals and precursors (on a factory basis and regarding those facilities producing more than one metric tonne annually). Whereas for medium-risk chemicals and precursors, any facility producing more than 20 tonnes annually was to be reported, together with the location of the facility and its maximum design capacity. For this category, annual reports on the production and the aggregate export and import quantities of each substance were deemed necessary. The UK approach was strongly supported by the other Western states in the CD negotiations.

Nevertheless, the concept of mandatory regular international inspection within the civilian chemical industries of States Parties, even when based on a random-selection procedure and limited to certain sensitive facilities, was not accepted by the group of socialist countries. Their approach continued to be based on national verification and international data exchange, combined with possible on-challenge inspections. Basic arguments presented in support of this approach have been:

- International verification (even at random) of facilities within the civilian chemical industry on any systematic (routine) scale poses an unacceptable burden to the economic and developmental aims of the future States Parties.
- The major aim of the international verification regime of the CW Convention is to supervise the implementation of the obligations the States Parties have accepted to undergo, not to control their civilian industry which is not connected to preparations for chemical warfare.
- Under the present technical conditions of binary technology, the proposed international mechanism will fail to assure that no clandestine CW production has been started.
- Given the existing scientific, technical and commercial links between states in the field of civilian chemical production, any significant production of chemical weapons will be detected even without the proposed international procedure.

Finally, some remarks on the implications of the draft US Chemical Weapons Convention (CD/500) for verification of non-production of chemical weapons. For civilian chemical enterprises, it calls for either ´special inspections´ (under the ´open invitation´) or routine random inspection depending on the ownership of or control over a certain facility, and for *ad hoc* (on challenge) inspections. For those chemicals compiled in a specific list (list C in the annex to the Convention), periodic inspections and annual reporting were foreseen. Inspections should include interviews, optical assessment of agreed areas, sampling at agreed points with subsequent analysis and the application of autonomous samplers at certain points (final products). Plans to change the final product of the facility or the production technology were to be reported in advance. The areas in the chemical industries covered by these procedures were the following:

-chemicals containing a methyl-phosphorus, ethyl-phosphorus or propyl-phosphorus bond;
-methyl and/or ethyl esters of phosphorous acid;
-3,3-dimethyl-butanol-2 (pinacolyl alcohol);
-N,N-disubstituted beta-aminoethanol (-ethane thiol, -halides);
-phenyl-, alkyl-, or cycloakyl-substituted glycollic acids; and
-4- or 3-hydroxy piperidine and their derivatives.

For dual-purpose agents (phosgene, cyanogen chloride, hydrogen cyanide, phosphorus trichloride, phosphorus oxychloride, chlorpicrine, thiodiglycol), an annual reporting to the Consultative Committee was proposed.

As discussed earlier, the major disagreement with respect to the US draft concerned the 'open invitation' concept.

In 1985 the discussion on the verification issue continued, with some fresh approaches being proposed but with no overall agreement being reached. The issues of verification within the chemical industry received some more attention as a result of these discussions

One important area associated with this debate is that of the categorization of chemicals for verification purposes. The original British approach of using a risk criterion (both in terms of toxicity and of the overall hazard to the Convention) was taken up and modified by the Swedish delegation (CD/632). The criteria applied were: the acute toxicity of a chemical, its earlier use as a chemical-warfare agent and its civilian application (including quantity considerations). This approach would lead to the creation of different lists with distinct verification requirements, and there would be a need for agreed procedures to establish these lists and to move a chemical from one list to another in cases where certain characteristics of the chemical change (for example, its civilian utility).

Comparable considerations were tabled by other countries, under the heading of permitted activities. A Yugoslavian paper (CD/613) discussed the problem of permitted activities and also considered the production of large-scale chemicals within the civilian chemical industry. The proposed approach comprised declarations, notifications of transport (which was defined as transfer to other facilities or export) and national verification combined with international on-site inspections on challenge.

With respect to highly-toxic civilian chemicals, such as certain pharmaceuticals, it was proposed that these be produced in small-scale facilities (the number of which would depend on the developmental stage of the pharmaceutical industry of a given country) with the application of a quantity criterion. Such facilities would be equipped with monitoring devices and automation, in essentially the same way as proposed for the small-scale facility for production for protective purposes. The latter would still exist and be operated under stringent international control. This approach ´uncoupled´ the production of warfare chemicals for protective purposes from that of other supertoxic chemicals which might be needed for certain peaceful, non-protective aims (such as certain pharmaceuticals).

The same problem was briefly addressed in a Finnish working paper (CD/CW/WP/92) which concentrated mainly on the verification requirements and possibilities in a small-scale facility (without further detailing the purpose of this facility, either for permitted or protective purposes). A description of a possible design of such a facility was presented, and techniques for monitoring this potential facility were elaborated in detail.

The fundamental disagreement regarding verification in the chemical industry, namely, national verification and international on-site inspection on challenge vs. random routine international on-site inspections, could not be resolved in 1985.

The two positions were further elaborated and detailed. The GDR delegation tabled a working paper (CD/620) explaining the concept of national verification and outlining the role and functions of such a national system for implementation of the Convention. Under this approach, the national system would comprise:

- the requisite laws and regulations to constitute a structure for implementing the Convention;
- a national authority (for control, co-operation with international bodies and on interstate levels, etc.); and

-organizational and operational elements at the facility level (to facilitate the function of the national, and--in the event of international on-challenge procedures--the international inspections).

The functions, rights and duties of the national authority were also specified. Together with the elements provided for in the verified facilities, this national authority would form the framework for the national systems of accounting and control of chemicals, which would be comprised of information (data generation and submission) and inspections. Under the GDR approach the national inspectors would be entitled to:

-visit all facilities involved in activities regarding the chemicals concerned (this wording is obviously not limited to production facilities);
-request information concerning the production and use of the monitored chemicals; and
-check all relevant facility records.

The approach of the Western state group was detailed in a FRG working paper (CD/627). One significant consideration in this document was that of limiting the scope of the requisite definitions for verification purposes within the civilian chemical industry. For the supertoxic lethal category, the toxicity criterion was supplemented by the assessment of the potential military threat of a certain chemical. ´Key precursors´ were also included under this category. This term was limited to those products in the final technical reaction stage, which are characteristic for the toxicity of the end-product.

Other precursors could be treated in the same way as key precursors if they posed a risk to the Convention, and if this could be agreed upon. In applying these criteria, the category of supertoxic lethal chemicals would actually not contain chemicals which are produced in the civilian chemical industry, but only those produced in declared, separate facilities for these purposes. The systems of international, routine, random inspections within the civilian chemical industry, would then be limited to control of the production of key precursors for the manufacture of supertoxic lethal chemicals. Further, an information exchange of data concerning the production and use of certain selected precursors (other than key precursors) should take place in order to facilitate the monitoring of particularly dangerous supertoxic lethal chemicals (examples are esters of phosphorus acid, phosphorus trichloride and phosphorous oxychloride).

Using mustard gas as an example, it was stressed that the scale of civilian production and application of a key precursor is of eminent importance for deciding the kind of verification. A restriction on the key precursor thiodiglycol would be insufficient to protect the aims of the Convention, but the inclusion of the other possible key precursors (sulphur chlorides) would be hardly practicable since these chemicals are produced and used in large quantities as basic substances in the chemical industry.

The verification of this latter case would therefore, under the FRG approach, be limited to data exchange. In addition, it was suggested that countries could exercise export controls on the relevant chemicals. Other lethal chemicals would be covered by reporting of statistical data. For large-scale civilian chemicals in this category it was seriously questioned if this data exchange would make any sense. With respect to the international routine inspections (for key precursors) inspections would concentrate on the quantity of chemicals produced or, where processing is continuous, on the output of finished products. It would not be necessary to install additional measuring instruments for the international inspections, and it was indicated that this might even be impracticable. The inspections would consist of a statistical-data survey, review of plant records, viewing of the facility areas, and sampling and analysis, but it would not provide a comprehensive survey of the facility and its activities.

In order to reduce the number of controlled facilities, only those receiving quantities of a certain key precursor larger than 1 metric tonne per year would become subject to routine international-control procedures.

With respect to other lethal chemicals, it was questioned how the international body envisaged under the Convention could determine from the total data submitted whether or not a state had violated the Convention, provided that there were data exchange of this category of chemicals. Moreover, the collection of comprehensive data must not lead to the disclosure of information about the relationships with buyers in a manner and to an extent that might impair national economic interests.

In general, therefore, this FRG paper approved the principle of routine international on-site inspections on a random-selection basis, but strongly stressed the need for narrow definitions regarding the chemicals which would fall under this control.

Another, related approach was offered by the Australian delegation in working paper CD/CW/WP/121. Chemicals were divided into three distinct categories, using the principal approach from CD/514 of the United Kingdom. The first category would cover those chemicals the production of which would actually be banned, such as the nerve agents and their key precursors, provided the latter have no civilian utility. The second category would contain supertoxic lethal chemicals with some peaceful application and key precursors with small-scale civilian use. Both would be subject to a stringent process of inspection, materials-accountancy and justification of use. The third category would cover the remaining relevant chemicals, such as ´dual purpose agents´ and other precursors, and would be subject to monitoring by data exchange and materials-accountancy. This third category would also include chemicals with large-scale civilian use.

In summary, the proposed approach was composed of:

- materials-accountancy (including quantity produced, end-user and purpose of production);
- routine, random inspections of chemical industry;
- import/export regulations and customs checks; and
- challenge inspections to resolve ambiguities.

There is, to sum up, a rather universal acknowledgement of the complications involved in the verification of non-production within the chemical industry of the future States Parties, and of the needs to:

- guarantee at least some verification (because of the possibilities, in principle, of misuse of this industrial branch for production of chemicals for warfare purposes); and
- protect the economic interests of this industry even under this projected verification regime.

One approach seems to attempt to solve this dilemma by narrowing the number of chemicals and facilities to be monitored, while still maintaining an international routine-inspection procedure. The other favours a rather comprehensive, but national system, combined with international data exchange and the provision for international on-challenge inspections.

V Summary

In reviewing the numerous proposals presented during the Geneva negotiations on chemical weapons, certain broadly agreed principles may be summarized as follows:

- Under a CW Convention, a Consultative Committee would be established, supported by an international secretariat, and would be responsible, *inter alia*, for the evaluation of regular reports submitted by the governments of the States Parties on the production and consumption of certain sensitive chemicals, including production data for their chemical industry, export and import data.
- The Consultative Committee would be in charge of carrying out inquiries and investigations, including necessary on-site inspections.
- The States Parties to the Convention would establish their own procedures for national supervision of the implementation of the obligations the Convention covers, for example, by establishing a national-verification agency or via other appropriate measures. They would collect and analyse data, and thereafter report to the Consultative Committee as indicated above.
- The verification regime would be different with respect to the risk a given chemical poses to the Convention, both in terms of toxicity and of its role in potential CW production. Certain chemicals would be prohibited for production except for permitted purposes in an aggregate annual quantity possessed by a party of one metric tonne, in one specialized small-scale facility per Party and under systematic international supervision. Other chemicals, including certain sensitive precursors, would be reported annually by national authorities and be

subject to at least international on-challenge procedures.

There are, however, certain unresolved areas as well:

- It has not been agreed whether systematic inspection of the chemical industries (or parts of them) of the States Parties would be performed by national or by international authorities.
- It has not been agreed whether a challenge for an *ad hoc* international inspection would be mandatory, or if a country would possess the right to reject on-site inspections if it were to adequately explain this decision.
- There has been complete disagreement with respect to ´special investigations´ as proposed in the US draft CW Convention CD/500.
- Finally, there has been at least some divergence with respect to chemicals falling, by their toxicity, under the category ´supertoxic lethal chemicals´ but having civilian application. The Western approach has called for their inclusion in systematic random-inspections on site. The approach of the socialist countries has seemed, at least implicitly, to tend to include these chemicals in the internationally controlled small-scale facility with a strictly limited production output.

NOTES AND REFERENCES

1. For States Parties to the 1925 Geneva Protocol, see Stockholm International Peace Research Institute (SIPRI), World Armaments and Disarmament: SIPRI Yearbook 1985 (Taylor & Francis: London, 1985).

2. For the text of reservations to the 1925 Geneva Protocol see, for example, Stockholm International Peace Research Institute (SIPRI), World Armaments and Disarmament: SIPRI Yearbook 1984 (Taylor & Francis: London, 1984).

3. For States Parties to the 1975 Biological and Toxin Weapons Convention, see SIPRI Yearbook 1985.

4. Argentina, Brazil, Burma, Egypt, Ethiopia, Mexico, Morocco, Nigeria, Sweden, Yugoslavia, CCD Working Document, 'Working paper on the prohibition of the development, production and stockpiling of chemical weapons and on their destruction', CCD/400, 26 Apr. 1973.

5. Japan, CCD Working Document, 'Draft convention on the prohibition of the development, production and stockpiling of chemical weapons and on their destruction', CCD/420, 30 Apr. 1974.

6. United Kingdom, CCD Working Document, 'Draft convention on the prohibition of the development, production and stockpiling of chemical weapons and on their destruction', CCD/512 and Corr.1., 6 Aug. 1976.

7. USSR and USA, CD Working Document, 'USSR-United States joint report on progress in the bilateral negotiations on the prohibition of chemical weapons', CD/48, 7 Aug. 1979.

8. Stockholm International Peace Research Institute (SIPRI), World Armaments and Disarmament: SIPRI Yearbook 1980 (Taylor & Francis: London, 1980), p. 368.

9. USSR and USA, CD Working Document, 'USSR-United States joint report on the progress in the bilateral negotiations on the prohibition of

chemical weapons', CD/112, 7 July 1980.

10. *Ad Hoc* Working Group on Chemical Weapons, CD/539, 28 Aug. 1984.

11. USA, 'Speech of the US ambassador to the CD', CD/PV.303, 28 Mar. 1985.

12. Quoted from *SIPRI Yearbook 1985*, p. 466.

13. USA, CCD Working Document, 'Working paper on economic data monitoring as a means of verifying compliance with a ban on chemical weapons', CCD/311, 15 Aug. 1970.

14. USSR, CCD Working Document, 'Working paper on the complete prohibition of chemical and bacteriological weapons', CCD/303, 6 Aug. 1970.

15. Bulgaria, Czechoslovakia, Hungary, Mongolia, Poland, Romania, USSR, CCD Working Paper, 'Working paper on ways of implementing control over compliance with a convention on the prohibition of the development, production and stockpiling of chemical weapons and on their destruction', CCD/403, 28 June 1973.

16. USA, CCD Working Document CCD/311 (see note 13), and 'Working paper on statistics relating to production and trade of certain chemical substances in the United States', CCD/368, 20 June 1972.

17. France, CD Working Document, 'Control of non-manufacture and non-possession of agents and weapons of chemical warfare', CD/106, 22 June 1980.

18. USSR, 'Speech of the Soviet ambassador to the CD', CD/PV.202, 10 Mar. 1983.

19. UK, CD Working Document, 'Verification of non-production of chemical weapons', CD/514, 10 July 1984.

20. FRG, CD/CW Conference room paper CD/CW/CRP.90, 'Concerning chemicals containing the methyl-phosphorus bond', 26 Jan. 1984.

21. FRG, CD Working Document, ´Working paper on some aspects of international verification of the non-production of chemical weapons: experience gained in the Federal Republic of Germany´, CD/37, 12 July 1979.

22. FRG, CD Working Document, ´Working paper on principles and rules for verifying compliance with a chemical weapons convention´, CD/265, 24 Mar. 1982.

23. UK, CD Working Document, ´Verification of non-production of chemical weapons´, CD/353, 8 Mar. 1983.

24. UK, CD Working Document, ´Verification of non-production of chemical weapons: proposals for inspection procedures and data exchange´, CD/575, 6 Mar. 1985.

25. UK, CD Working Document, ´Chemical Weapons Convention: the organs and constitution of the organization´, CD/589, 11 Apr. 1985.

APPENDIX I. PROPOSALS FOR GENERAL PRINCIPLES REGARDING THE VERIFICATION OF COMPLIANCE WITH A CONVENTION BANNING CHEMICAL WEAPONS

Complaint procedure with the UN Security Council

Canada (CCD 300), Czechoslovakia (CCD 299), Japan (CCD 301), Socialist States Group (CCD 361), UAR/Egypt (CCD 314), USA (CD 48, CD 112, CD 343, CD 500), USSR (CDD 303, CD 48, CD 112, CD 294), Yugoslavia (CCD 302, CCD 377)

Establishment of a national body to monitor the implementation and to verify compliance

Canada (CD 167), Japan (CCD 420), Socialist States Group (CCD 403), USA (CCD 437, implicitly), USSR (CCD 403, CD 294)

Establishment of an international organ responsible, *inter alia*, for verification purposes (International Verification Agency, Consultative Committee, and so on)

Canada (CD 167, CD 313), China (CD 102, CD 212, CD 443), FR Germany (CD 37, CD 265, CD 326), Japan (CCD 413, CCD 420), the Netherlands (CD 203, CD 445), Socialist States Group (CD 532), Sweden (CCD 322), USA (CCD 437, CD 48, CD 112, CD 343, CD 500), USSR (CD 48, CD 112, CD 294, CD 532)

Combination of national and international verification means

Canada (CD 167), France (CD 106, implicitly), FR Germany (CD 326), Japan (CCD 420, CCD 430), Sweden (CCD 322), UK (CD 244), USA (CCD 437, CD 48, CD 112), USSR (CD 48, CD 112, CD 294), Yugoslavia (CCD 377, CD 482)

Ad hoc (by challenge) international inspections, in general

FR Germany (CD 37), Japan (CCD 301, CCD 413, CCD 420), Sweden (CCD 322), UK (CCD 512, CD 244), USA (CD 48, CD 112, CD 343, CD 500), Yugoslavia (CCD 302, CCD 377)

Regular inspections in the chemical industry on a quota basis

UK (CCD 512, CD 15)

Regular international inspections in the chemical industry on a random basis

China (CD 443), FR Germany (CD 265, CD 326), the Netherlands (CD 445), Norway (CD 397), UK (CD 244, CD 353, CD 514, CD 575, CD 589), USA (CD 500)

National statistical surveillance of certain civilian chemical production and reporting to an international body

Canada (CD 167), Czechoslovakia (CCD 299), Egypt/UAR (CCD 314), France (CD 106), Italy (CCD 335, CCD 373), Japan (CCD 288, CCD 413), Socialist States Group (CCD 361, CCD 403), Sweden (CCD 322), USA (CD 48, CD 112), USSR (CCD 303, CCD 361, CCD 403, CD 48, CD 294), Yugoslavia (CCD 302)

Avoidance of hampering the economic and technological developments of the States Parties

Japan (CCD 420), Socialist States Group (CCD 361), UK (CCD 512), USA (CD 48, CD 112), USSR (CD 48, CD 112, CD 294)

Special investigations (based on ´open invitations´) in governmental owned or controlled facilities

USA (CD 500)

APPENDIX II. PROPOSED PROCEDURES TO VERIFY, WITHIN THE FRAMEWORK OF AN INTERNATIONAL BAN ON CHEMICAL WEAPONS, CERTAIN ACTIVITIES IN THE CIVILIAN CHEMICAL INDUSTRY

Statistics on sensitive material flows (by national bodies)

China (CD 443), France (CD 106), FR Germany (CD 265), Italy (CCD 335, CCD 373), Japan (CCD 288, CCD 301, CCD 413, CCD 420), Socialist States Group (CCD 403), Sweden (CCD 322), UK (CD 353), USA (CD 48, but having pointed to strong limitations earlier, in CCD 293, CCD 311, CCD 368), USSR (CCD 403, CD 48), Yugoslavia (CCD 302, CCD 482)

Control of authorized material handling for certain high-risk chemicals (by international body)

USA (CCD 437, CD 500)

Check of safety precautions and technical inspections (by international body)

FR Germany (CD 37, CD 326), UK (CD 353, CD 575) USA (CD 500, implicitly)

Near-site (for example, downstream) sampling and analysis (by international body)

France (CD 106), FR Germany (CD 265), the Netherlands (CCD 533, CD 306, CD 307), UK (CD 244)

Extraterritorial surveillance (by national bodies)

USSR (CCD 538)

Examination of research work in the chemical industry (by national bodies)

Socialist States Group (CCD 403)

Prohibition of patenting of means of chemical warfare (agents, delivery systems, special equipment except protective materials) (by national bodies)

Socialist States Group (CCD 403)

Sampling and analysis of enterprise materials (raw materials, streamline material, intermediates, final products) (by international body)

Canada (CD 167), FR Germany (CD 265, CD 326), UK (CD 575), USA (CD 500)

Interviews (by international body)

USA (CD 500)

Autonomous sampling of final products in a continuous way (by international body)

USA (CD 500)

Examination of enterprise records (by international body)

UK (CD 575), USA (CD 500)

Photographic documentation (by international body)

USA (CD 500)

Declaration of location, capacity, basic technical/technological data and the like for plants potentially capable of producing CW agents

China (CD 443), UK (CD 514, CD 575)

APPENDIX III. AREAS WITHIN THE CIVILIAN CHEMICAL INDUSTRY WHICH MIGHT BE SUBJECT TO VERIFICATION MEASURES

1. Facilities processing phosphorus chemicals

A. Facilities processing or producing elementary phosphorus, phosphorus trichloride, phosphorus oxychloride, phosphorus pentasulfide, phosphorus pentachloride

Tracing of material flows (production, import, export, stockpiling, end use)

Japan (CCD 288)

PCl_3 and $POCl_3$: statistical monitoring

Italy (CCD 335)

Regular reporting to include information/data exchange on production statistics, for PCl_3 and $POCl_3$

UK (CD 514)

PCl_3 and $POCl_3$: reporting annually the location of each production facility and statistical data on the aggregate quantities produced, imported and exported, and on the end uses

USA (CD 500)

B. Facilities producing organophosphorus chemicals

B.1. In general

International random on-site inspections including optical evaluation, sampling and analysis

FR Germany (CD 37, CD 265, CD 326), UK (CCD 512)

Brief international on-site inspections, combined with effluent analysis

France (CD 106)

Downstream water analysis to detect methyl-phosphorus chemicals

The Netherlands (CCD 533, CD 306, CD 307)

B.2. Facilities producing alkyl-phosphorus chemicals, where alkyl is methyl, ethyl or propyl

Ban on any production of methyl-phosphorus chemicals

USSR (CD/Pv.202), Australia (CD/CW/WP.25, for a period ending 10 years after entry into force of the Convention)

Annual reporting for each chemical produced, imported or exported in an aggregate amount greater than a certain limit, the location of each facility, the respective aggregate quantities, and the end uses. Plans to establish a new production facility or to significantly change the capacity of an existing one should be reported 90 days in advance. Systematic international on-site inspections

USA (CD 500)

Regular reporting which will include description/justification of the civil uses for which the chemicals are produced plus routine random-selected international on-site inspections

UK (CD 514)

B.3. Facilities producing Di- or Trimethylesters of phosphorous acid (P-III)

Annual reporting for any chemical produced, imported and exported in an aggregate quantity greater than ..., the location of the respective facility and statistical data on the aggregate production, import and export, and the end uses of the chemical. Plans to establish a new production facility or to substantially change the capacity of a facility should be reported 90 days in advance. Subject to international systematic on-site inspections

USA (CD 500)

Regular reporting which will include description/justification of the civil uses of the chemical. Routine, random on-site international inspections

UK (CD 514)

Tracing of material flows

Japan (CCD 288)

B.4. Facilities producing methylphosphonic acid--difluoride

Only limited production, under systematic international control and within an aggregate quantity annually produced or retained of other supertoxic lethal chemicals and their key precursors of one metric tonne per state, in a single facility per state

USSR (CD 294)

As under B.2

UK (CD 514)

Information on persons authorized to possess such chemicals, the quantity produced and used at each location and the end uses should be reported annually. Production at a single specialized facility per party and within an aggregate annual quantity produced and retained of supertoxic chemicals and their precursors not exceeding one metric tonne (?). Systematic international on-site inspections

USA (CD 500)

2. Facilities producing chemicals exceeding a certain toxicity limit

Limitation of the production for each State Party to one single specialized facility, with an aggregate annual quantity produced and retained not exceeding one metric tonne for all relevant chemicals and their key precursors. Systematic international control

USSR/USA (CD 48, CD 112)
USA (CD 500) for defensive purposes
USSR (CD 294) for all permitted purposes

Statistical reporting on a factory base

Japan (CCD 301)

National as well as international verification (toxicity limit at that time set to LD_{50} lower 1 mg kg^{-1})

Sweden (CCD 322)

Production under international supervision

Yugoslavia (CD 482)

Information on the persons authorized to possess such chemicals, the quantity produced and used at each location and the end uses to be reported annually

USA (CD 500)

Regular reporting which will include description/justification of the civil use for which the chemical is produced, and routine random international on-site inspection

UK (CD 514)

3. Facilities producing other chemicals

A. Arsenic trichloride

Regular reporting, routine international on-site inspections

UK (CD 514)

B. Hydrogen cyanide, phosgene, cyanogen chloride

Regular reporting and data exchange

China (CD 443), Italy (CCD 373), Japan (CCD 301), UK (CD 514), USA (CD 500), Yugoslavia (CD 482)

C. Pinacolyl alcohol

Regular reporting and data exchange, random international on-site inspections

UK (CD 514), USA (CD 500)

D. N,N-Diisopropylaminoethyl-2-halides (-ol, -thiol)

Regular reporting and data exchange, random international on-site inspections

UK (CD 514)

E. Aryl- (alkyl-, cycloalkyl) glycolic acids-esters

as D

UK (CD 514)

Regular reporting plus systematic international on-site inspections (for Phenyl-, alkyl- or cycloalkyl substituted glycolic acids)

USA (CD 500)

F. 2,2´-dihydroxyethylsulphide (Thiodiglycol)

as D

UK (CD 514), USA (CD 500)

G. N, N-disubstituted - beta - aminoethanols (-halides, -thiols)

Regular reporting and data exchange

UK (CD 514)

Regular reporting plus systematic international on-site inspections

USA (CD 500)

H. Sulphur monochloride (S_2Cl_2)

Regular reporting and data exchange

UK (CD 514)

J. 3- or 4-hydroxy piperidine and their derivatives

Regular reporting and data exchange plus systematic international on-site inspections

USA (CD 500)

Regular reporting and data exchange

UK (CD 514)

K. Trichloronitromethane (chloropicrin)

as D

USA (CD 500)

SIPRI publications on CBW

- The Problem of Chemical and Biological Warfare. Vol. 1–6:
  1. The rise of CB weapons. 1971. ISBN 91-85114-10-3.
  2. CB weapons today. 1973. 91-85114-16-2.
  3. CBW and the law of war. 1974. 91-85114-17-0.
  4. CB disarmament negotiations, 1920–1970. 1971. 91-85114-11-1.
  5. The prevention of CBW. 1971. 91-85114-13-8.
  6. Technical aspects of early warning and verification. 1975. 91-85114-18-9.

- Chapters in the SIPRI Yearbooks:

  SIPRI yearbook of world armaments and disarmament, 1968/69. 1969. 91-85114-03-0.
  pp. 112-34. Development in chemical and biological warfare.
  ——— 1969/70. 1970. 91-85114-07-3.
  pp. 185-206. The CBW debate and other disarmament measures.
  World armaments and disarmament, SIPRI yearbook 1972. 1972. 91-85114-12-X.
  pp. 501-22. Chemical and biological disarmament.
  ——— 1973. 91-85114-19-7.
  pp. 383-91. Chemical disarmament.
  ——— 1974. 0-262-19129-6.
  pp. 370-84. Chemical disarmament.
  ——— 1975. 0-262-19140-7.
  pp. 426-32. Chemical disarmament.
  ——— 1977. 91-22-00116-6.
  pp. 86-102. Dioxin: a potential chemical-warfare agent; and
  pp. 364-67. Prohibition of biological and chemical weapons.
  ——— 1978. 0-85066-134-X.
  pp. 360-76. The destruction of chemical warfare agents.
  ——— 1979. 0-85066-181-1.
  pp. 470-89. Stockpiles of chemical weapons and their destruction.
  ——— 1980. 0-85066-201-X.
  pp. 365-79. Chemical disarmament.
  ——— 1982. 0-85066-230-3.
  pp. 317-61. The changing status of chemical and biological warfare: recent technical, military and political developments; and
  pp. 456-57. Chemical weapons.
  ——— 1983. 0-85066-247-8.
  pp. 391-426. Chemical and biological warfare: developments in 1982; and
  pp. 563-67. Chemical disarmament.
  ——— 1984. 0-85066-263-X.
  pp. 319-49. Chemical and biological warfare: developments in 1983; and
  pp. 421-54. Implications of genetic engineering for chemical and biological warfare.
  ——— 1985. 0-85066-297-4.
  pp. 159-205. Chemical and biological warfare: developments in 1984; and
  pp. 206-219. An analysis of the reports of Iraqi chemical warfare against Iran, 1980–84.

- Other SIPRI publications:

  Chemical disarmament: some problems of verification. 91-85114-20-0.

  The effects of developments in the biological and chemical sciences on CW disarmament negotiations, SIPRI research report 13. SIPRI, 1974.

  Delayed toxic effects of chemical warfare agents. 91-85114-29-4.

  Chemical disarmament: new weapons for old. 91-85114-27-8.

  Medical protection against chemical-warfare agents. 91-22000-44-5.

  Ecological consequences of the Second Indochina War. 91-22000-62-3. pp. 24–45, 53–55.

  Weapons of mass destruction and the environment. 0-85066-132-3. pp. 31–48.

  The fight against infectious diseases: a role for applied microbiology in military redeployment. SIPRI, 1979, mimeo. 91-85114-26-X.

  Chemical weapons: destruction and conversion. 0-85066-199-4.

  Herbicides in war. 0-85066-265-6.

SIPRI publications may be ordered from SIPRI.

The manufacturer's authorised representative in the EU for product safety is Oxford University Press España S.A. of el Parque Empresarial San Fernando de Henares, Avenida de Castilla, 2 – 28830 Madrid (www.oup.es/en or product.safety@oup.com). OUP España S.A. also acts as importer into Spain of products made by the manufacturer.

www.ingramcontent.com/pod-product-compliance
Lightning Source LLC
LaVergne TN
LVHW060212210926
848481LV00008B/2708
9780198291077